Thomas R. Gulledge   William P. Hutzler
Joan S. Lovelace   Editors

Cost Estimating and Analysis

Balancing Technology and Declining Budgets

With 48 Figures

Springer-Verlag

New York Berlin Heidelberg London Paris
Tokyo Hong Kong Barcelona Budapest

Thomas R. Gulledge
The Institute of Public Policy
George Mason University
Fairfax, VA 22030-4444
USA

William P. Hutzler
Economic Analysis Center
The MITRE Corporation
McLean, VA 22102-3481
USA

Joan S. Lovelace
Economic Analysis Center
The MITRE Corporation
McLean, VA 22102-3481
USA

Library of Congress Cataloging-in-Publication Data
Gulledge, Thomas R., 1947–
    Cost estimating and analysis : balancing technology and declining
budgets / Thomas R. Gulledge, William P. Hutzler, Joan S. Lovelace, editors.
       p.   cm.
    Includes bibliographical references.
    ISBN 0-387-97838-0. — ISBN 3-540-97838-0
    1. Cost effectiveness.   2. Costs, Industrial.   3. Value analysis
(Cost Control).   4. Operations research.   I. Hutzler, William P.
II. Lovelace, Joan S.   III. Title.
HD47.4.G85   1992
658.15′52—dc20   92-9961

Printed on acid-free paper.

Production managed by Hal Henglein; manufacturing supervised by Robert Paella.
Camera-ready copy prepared by the contributors.
Printed and bound by Edwards Brothers, Inc., Ann Arbor, MI.
Printed in the United States of America.

9 8 7 6 5 4 3 2 1

ISBN 0-387-97838-0 Springer-Verlag New York Berlin Heidelberg
ISBN 3-540-97838-0 Springer-Verlag Berlin Heidelberg New York

Preface

This volume presents a selection of the presentations from the annual conference of the Society of Cost Estimating and Analysis that was held in McLean, Virginia in the summer of 1992. The papers are representative of the issues that are of interest to researchers in the cost analysis community.

Cost analysis is interdisciplinary. Its foundations are in the intersection of operations research, managerial accounting, systems analysis, and economics. These papers are representative of the diversity of our discipline. The topics included in this volume cover both public and private sector issues, and they range from manufacturing systems analysis to software engineering economics. In all cases, the emphasis is on modeling, understanding, and predicting the costs, benefits, and risks associated with alternative actions within organizations.

This is a particularly challenging time for the cost analysis profession; hence the conference theme: *Balancing Technology and Declining Budgets*. With government budgets declining and the concurrent reengineering of public and private organizations, there is much demand for systems analysis. New methodologies are needed to model the productivity enhancing effects of investments in new technologies, while simultaneously quantifying the cost savings from implementing new business processes. These issues were addressed at our annual conference, and they will continue to be the important research topics at future conferences.

We thank everyone who helped make this conference a success, especially those who graciously allowed us to include their work in this volume.

Thomas R. Gulledge
The Institute of Public Policy
George Mason University
Fairfax, Virginia

William P. Hutzler and Joan S. Lovelace
Economic Analysis Center
The MITRE Corporation
McLean, Virginia

Contents

IV. Systems Cost Analysis

V. Cost and Production Analysis

VI. Cost Sensitivity Analysis

I. Learning and Production Rate

Learning Curve and Rate Adjustment Models: An Investigation of Bias

O. Douglas Moses
Department of Administrative Science, Naval Postgraduate School,
Monterey, CA 93943

ABSTRACT

Learning curve models have gained widespread acceptance as a technique for analyzing and forecasting the cost of items produced from a repetitive process. Considerable research has investigated augmenting the traditional learning curve model with the addition of a production rate variable, creating a rate adjustment model. This study compares the forecasting bias of tne learning curve and rate adjustment models. A simulation methodology is used to vary conditions along seven dimensions. The magnitude and direction of errors in estimating future cost are analyzed and compared under the various simulated conditions, using ANOVA. Overall results indicate that the rate adjustment model is generally unbiased. If the cost item being forecast contains any element that is not subject to learning then the traditional learning curve model is consistently biased toward underestimation of future cost. Conditions when the bias is strongest are identified.[1]

INTRODUCTION

The problem of cost overruns has consistently plagued the process of acquiring weapons systems by the U. S. Department of Defense. Technical improvements in the conduct of cost estimation and institutional changes in the process of procurement have occurred over the past few decades, but unanticipated cost growth during procurement continues. A cost overrun, by definition, occurs when the actual cost of a program exceeds the estimated cost. There are, in principle, two broad reasons that a cost overrun could occur. Either (a) initial cost estimates are fair when made, but

---

[1]This paper was prepared for the Cost Estimating and Analysis Division of the Naval Sea Systems Command. Funding was provided by the Naval Postgraduate School. This paper is a companion to an earlier paper [21]. Both papers investigate and evaluate two cost estimating approaches commonly used by cost analysts. Both use the same methodology. The earlier paper focused on investigating the accuracy of the two approaches; the current paper focuses on bias. Readers familiar with the earlier paper will find the first 11 pages of this paper, describing the methodology, to be quite familiar. For readers unfamiliar the the earlier paper, this current paper is designed to be self-contained.

subsequently actual costs are poorly managed and controlled; or (b) actual costs are well managed, but initial cost estimates were unrealistic. This paper focuses on the latter situation. The paper examines and compares bias in two estimating models used frequently by cost analysts: the learning curve and the rate adjustment model.

Learning curves have gained widespread acceptance as a tool for planning, analyzing, explaining, and predicting the behavior of the unit cost of items produced from a repetitive production process. (See Yelle [31] for a review.) Cost estimation techniques for planning the cost of acquiring weapon systems by the Department of Defense, for example, typically consider the role of learning in the estimation process. The premise of learning curve analysis is that cumulative quantity is the primary driver of unit cost. Unit cost is expected to decline as cumulative quantity increases.

There is general acknowledgement that cumulative quantity is not the only factor that influences unit cost and that the simple learning curve is not a fully adequate description of cost behavior. Hence prior research has attempted to augment learning curve models by including additional variables [e.g., 20]. Most attention has been focused on the addition of a production rate term.[2] The resulting augmented model is usually referred to as a rate adjustment model.

Conceptually, production rate should be expected to affect unit cost because of the impact of economies of scale. Higher production rates may lead to several related effects: greater specialization of labor, quantity discounts and efficiencies associated with raw materials purchases, and greater use of facilities, permitting fixed overhead costs to be spread over a larger output quantity. Together, these effects work to increase efficiency and reduce production cost [5, 6, 15, 18]. However, higher production rate does not guarantee lower cost. When production rate exceeds capacity, such factors as over-time pay, lack of skilled labor, or the need to bring more facilities online may lead to inefficiencies and increased unit cost. In short, production rate may be associated with both economies and diseconomies of scale.

PRIOR RESEARCH

Numerous studies, using data on actual production cost elements, have been conducted to empirically examine the impact of production rate on unit

---

[2]One review [7] of the literature pertaining to learning curves found that 36% of the articles reviewed attempted to augment the learning curve model in some manner by the inclusion of production related variables.

cost. The broad objective of the research has been to document rate/cost relationships and determine if consideration of production rate leads to improvements in cost explanation or prediction. Results have been inconsistent and general findings inconclusive. Various studies [1, 8, 13, 14] found little or no significance for rate variables. Other studies did document significant rate/cost relationships [5, 10]. Some research found significant results only for particular individual cost elements, such as labor [25], tooling [16] or overhead [15]. But rate/cost relationships for these same cost elements were not consistently evident in other studies. When significant, estimates of the rate/cost slope varied greatly and the direction of the relationship was sometimes negative and sometimes positive [e.g., 20]. In reviewing the existing research on production rate, Smith [23] concluded that a rate/cost relationship may exist but that the existence, strength and nature of the relationship varies with the item produced and the cost element examined.[3]

The prior research suggests that consideration of production rate sometimes improves cost explanation, but not always. The prior research suggests that a traditional learning curve model sometimes is preferable to a rate adjustment model, but not always. The prior research provides little guidance concerning the circumstances under which explicit incorporation of production rate into a learning curve model is likely to lead to improved explanation or prediction. This issue is important in a number of cost analysis and cost estimation situations. Dorsett [11], for example, describes the current situation facing military cost estimators who, with the military facing budget reductions and program stretchouts, are required to rapidly develop weapon system acquisition cost estimates under many different quantity profiles. One choice the cost analyst faces is between using a rate adjustment model or a traditional learning model to develop estimates.[4]

---

[3] Several explanations for these varying, inconclusive empirical results can be offered: (a) Varying results are to be expected because rate changes can lead to both economies and diseconomies of scale. (b) Production rate effects are difficult to isolate empirically because of colinearity with cumulative quantity [12]. (c) Researchers have usually used inappropriate measures of production rate leading to misspecified models [6]. (d) The impact of a production rate change is dominated by other uncertainties [15], particularly by cumulative quantity [2]. Alchian [1], for example, was unable to find results for rate adjustment models that improved on the traditional learning curve without a rate parameter.

[4]Two other techniques for making cost estimates when production rate changes are also mentioned by Dorsett: curve rotation, which involves and ad hoc upward or downward revision to the slope of the learning curve, and the use of repricing models [e.g., 3, 4] which adjust learning curve estimates to

Reacting to the inconsistent findings in the literature, Moses [21] raised the question of under what circumstances it would be beneficial to incorporate consideration of production rate into a cost estimation problem. The objective of the research was to attempt to identify conditions when a rate adjustment model would outperform the traditional learning curve model (and vice versa). The ability of each model to accurately estimate future cost was assessed under various conditions. Generally findings were that neither model dominated; each was relatively more accurate under certain conditions.

OBJECTIVE OF THE STUDY

One limitation of the Moses study was that accuracy was measured as the absolute difference between estimated and actual cost, without concern for the direction of the difference. When controlling real-world projects, the consequences of errors in estimation typically depend on whether costs are under or overestimated. Underestimation, resulting in cost growth or cost overruns, is typically met with less pleasure than overestimation. Thus the question of model bias toward over or underestimation is of interest.

The objective of this study is to investigate and compare estimation bias for the learning curve and rate adjustment models. Does either model exhibit consistent or systematic bias? Are there circumstances where one model may be biased and the other not? Is the bias produced toward underestimation or overestimation of future cost?

RESEARCH APPROACH

Operationally the research questions require an examination of the estimation errors from two competing cost estimation models. The two competing models were as follows:

The traditional learning curve model, which predicts unit cost as a function of cumulative quantity[5]:

---

reflect a greater or lesser application of overhead cost. Dorsett criticized curve rotation for being subjective and leading to a compounding of error when the prediction horizon is not short. He criticized repricing models because they must be plant-specific to be effective.

[5]Note that this is an incremental unit cost model rather than a cumulative average cost model. Liao [17] discusses the differences between the two approaches and discusses why the incremental model has become dominant in practice. One reason is that the cumulative model weights early observations more heavily and, in effect, "smooths" away period-to-period changes in average cost.

$$C_L = aQ^b \qquad (1)$$

where

C_L = Unit cost of item at quantity Q (i.e., with learning considered).
Q = Cumulative quantity produced.
a = Theoretical first unit cost.
b = Learning curve exponent (which can be converted to a learning slope by slope = 2^b).

And the most widely used rate adjustment model, which modifies the traditional learning curve model with the addition of a production rate term:

$$C_R = aQ^bR^c \qquad (2)$$

where

C_R = Unit cost of item at quantity Q and production rate per period R (i. e., with production rate as well as learning considered).
Q = Cumulative quantity produced.
R = Production rate per period measure.
a = Theoretical first unit cost.
b = Learning curve exponent.
c = Production rate exponent (which can be converted to a production rate slope by slope = 2^c).

A simulation approach was used to address the research questions. In brief, cost series were generated under varying simulated conditions. The learning curve model and the rate adjustment model were separately fit to the cost series to estimate model parameters. The estimated models were then used to separately predict future cost. Actual cost was compared with predicted cost to measure bias. Finally, an analysis (ANOVA) was conducted relating bias (dependent variable) to the simulated conditions (independent variables).

There are three main benefits gained from the simulation approach. First, factors hypothesized to influence bias can be varied over a wider range of conditions than would be encountered in any one (or many) sample(s) of actual cost data. Second, explicit control is achieved over the manipulation of factors. Third, noise caused by factors not explicitly investigated is removed. Hence simulation provides the most efficient way of investigating data containing a wide variety of combinations of the factor levels while controlling for the effects of other factors not explicitly identified.

RESEARCH CHOICES

There were five choices that had to be made in conducting the simulation experiment:

(1) The form of the rate adjustment (RA) model whose performance was to be compared to the learning curve (LC) model.

(2) The functional form of the cost model used to generate the simulated cost data.

(3) The conditions to be varied across simulation treatments.

(4) The cost objective (what cost was to be predicted).

(5) The measure of bias.

Items (1), (2), (4) and (5) deal with methodological issues. Item (3) deals with the various conditions simulated; conditions which may affect the nature and magnitude of bias. Each item will be discussed in turn.

    1. <u>The Rate Adjustment Model</u>. Various models, both theoretical and empirical, have been suggested for incorporating production rate into the learning curve [3, 4, 18, 23, 27, 29]. The models vary with respect to tradeoffs made between theoretical completeness and empirical tractability. Equation 2, described above, was the specific rate adjustment model analyzed in this study, for several reasons: First, it is the most widely used rate adjustment model in the published literature. Second, it is commonly used today in the practice of cost analysis [11]. Third, in addition to cost and quantity data (needed to estimate any LC model), equation 2 requires only production rate data.[6] Thus equation 2 is particularly appropriate for examining the incremental effect of attention to production rate. In short, equation 2 is the most widely applicable and most generally used rate adjustment model.

    2. <u>The Cost Generating Function</u>: A "true" cost function for an actual item depends on the item, the firm, the time period and all the varying circumstances surrounding actual production. It is likely that most manufacturers do not "know" the true cost function underlying goods they manufacture. Thus the choice of a cost function to generate simulated cost data is necessarily ad hoc. The objective here was to choose a "generic" cost function which had face validity, which included components (parameters and variables) that were generalizable to all production situations, and which resulted in a unit cost that depended on both learning and production

---

[6]Other RA models offered in the literature require knowledge of still additional variables. The equation 2 model is particularly applicable in situations where a cost analyst or estimator does not have ready access to or sufficient knowledge about the cost structure and cost drivers of a manufacturer. Examples include the Department of Defense procuring items from government contractors in the private sector, or prime contractors placing orders with subcontractors.

rate factors. The following explanation of the cost function used reflects these concerns.

At the most basic level the cost of any unit is just the sum of the variable cost directly incurred in creating the unit and the share of fixed costs assigned to the unit, where the amount of fixed costs assigned depend on the number of units produced.

$$UC = VC + \frac{FC}{PQ} \qquad (3)$$

where

UC = Unit cost.
VC = Variable cost per unit.
FC = Total fixed costs per period.
PQ = Production quantity per period.

The original concept of "learning" [30] involved the reduction in variable cost per unit expected with increases in cumulative quantity produced. (By definition, fixed costs are assumed to be unaffected by volume or quantity.) To incorporate the effect of learning, variable cost can be expressed as:

$$VC_Q = VC_1(Q^d) \qquad (4)$$

where

Q = Cumulative quantity.
VC_Q = Variable cost of the Qth unit.
VC_1 = Variable cost of the first unit.
d = Parameter, the learning index.

Substituting into equation 3:

$$UC_Q = VC_1(Q^d) + \frac{FC}{PQ} \qquad (5)$$

Additionally, assume the existence of a "standard" ("benchmark," "normal," "planned") production quantity per period (PQ_s). Standard fixed cost per unit (SFC) at the standard production quantity would be:

$$SFC = \frac{FC}{PQ_s} \qquad (6)$$

The production rate (PR) for any period can then be expressed as a ratio of the production quantity to the standard quantity:

$$PR = \frac{PQ}{PQ_s} \qquad (7)$$

The second term of equation (6) can then be rewritten as:

$$FC = \frac{SFC}{PR} \qquad (8)$$
$$\overline{PQ}$$

and equation 5 rewritten as:

$$UC_Q = VC_1(Q^d) + SFC(PR^{-1}) \qquad (9)$$

In this formulation it can be seen that total cost per unit is the sum of variable cost per unit (adjusted for learning) plus standard fixed cost per unit (adjusted for production rate). This model incorporates the two factors presumed to impact unit costs that have been most extensively investigated: cumulative quantity (Q) and production rate per period (PR).[7] It is consistent with both the theoretical and empirical literature which sees the primary impact of learning to be on variable costs and the primary impact of production rate to be on the spreading of fixed costs [23]. Simulated cost data in this study was generated using equation 9, while varying values for the variables and parameters on the right hand side of the equation to reflect differing conditions.

3. The Simulated Conditions: The general research hypothesis is that the estimation bias of the LC and RA models will depend on the circumstances in which they are used. What conditions might be hypothesized to affect bias? Seven different factors (independent variables) were varied during the simulation. These factors were selected for examination because they have been found to affect the magnitude of model prediction errors in prior research [21,26]. In the following paragraphs, each factor is discussed. A label for each is provided, along with a discussion of how the factor was operationalized in the simulation. Table 1 summarizes the seven factors.

i) Data History (DATAHIST): The number of data points available to estimate parameters for a model should affect the accuracy of a model. More data available during the model estimation period should be associated with greater accuracy for both the LC and the RA model.[8] The effect of the

---

[7]Smith [23, 24] for example, used a model similar to equation 9 to explore the effect of different production rates on unit cost. Balut [3] and Balut, Gulledge and Womer [4] construct models based on learning and production quantity to assist in "redistributing" overhead and "repricing" unit costs when changes in production rate occur. The Balut and Balut, Gulledge and Womer models differ in that they determine a learning rate for total (not variable) unit cost and then apply an adjustment factor to allow for the impact of varying production quantity on the amount of fixed cost included in total cost.

[8]There are, of course, cost/benefit tradeoffs. The marginal benefits of increased prediction accuracy for any model must be weighed against the marginal costs of additional data collection.

number of data points on bias however is unclear. If a model is inherently
an "incorrect," biased representation of a phenomena, having more data on

---

TABLE 1

INDEPENDANT VARIABLES

Concept	Label	Levels		
Data History	DATAHIST[a]	4	7	10
Variable Cost Learning Rate	VCRATE	75%	85%	95%
Fixed Cost Burden	BURDEN[b]	15%	33%	50%
Production Rate Trend	PROTREND[c]	Level	Growth	
Production Rate Instability/Variance	RATEVAR[d]	.05	.15	.25
Cost Noise/Variance	COSTVAR[e]	.05	.15	.25
Future Production Level	FUTUPROD[f]	Low	Same	High

   a.   Number of data points available during the model estimation period;
simulates the number of past production lots.
   b.   Standard per unit fixed cost as a percentage of cumulative average
per unit total cost, during the model estimation period.
   c.   A level trend means production at 100% of standard production for
each lot during the estimation period. A growth trend means production rate
gradually increasing to 100% of standard production during the estimation
period. The specific growth pattern depends on the number of production lots
in the estimation period, with sequences as follows (expressed as a % of 40%,
60%, 80%, 100%, 100%, 100%. For DATAHIST = 10: 10%, 20%, 35%, 50%, 70%, 90%,
100%, 100%, 100%, 100%.
   d.   Coefficient of variation of production rate. (Degree of instability
of production rate around the general production rate trend.)
   e.   Coefficient of variation of total per unit cost.
   f.   "Same" means production rate at 100% of standard for each lot
produced within the prediction zone. "Low" means production rate at 50%.
"High" means production rate at 150%.

---

which to estimate the model parameters will not eliminate the bias.

In the simulation, data history was varied from four to seven to ten data
points available for estimating model parameters. This simulates having
knowledge of costs and quantities for four, seven or ten production lots.
Four is the minimum number of observations needed to estimate the parameters
of the RA model by regression. The simulation focuses on lean data

availability both because the effects of marginal changes in data availability should be most pronounced when few observations are available and because many real world applications (e.g., cost analysis of Department of Defense weapon system procurement) occur under lean data conditions.

ii) Variable Cost Learning Rate (VCRATE): In the cost generating function, learning affects total unit cost by affecting variable cost per unit. Past research [26] has shown that the improvement in prediction accuracy from including a learning parameter in a model (when compared to its absence) depends on the degree of learning that exists in the underlying phenomena being modeled. The association between learning rate and degree of bias however is unclear. In the simulation, variable cost learning rate (reflected in parameter d in equation 9) was varied from 75% to 85% to 95%. Generally, complex products or labor intensive processes tend to experience high rates of learning (70-80%) while simple products or machine-paced processes experience lower (90-100%) rates [26]. [9]

iii) Fixed Cost Burden (BURDEN): In theory (and in the cost function, equation 9) a change in the number of units produced during a period affects unit cost in two ways: First, increasing volume increases cumulative quantity and decreases variable cost per unit, due to learning. Second, increasing volume increase the production rate for a period and reduces fixed cost per unit, due to the spreading of total fixed cost over a larger output. Both these effects operate in the same direction; i. e., increasing volume leads to lower per unit cost. This has led some cost analysts to conclude that in practice, it is sufficient to use an LC model, letting the cumulative quantity variable reflect the dual impacts of increased volume. Adding a production rate term to an LC model is seen as empirically unnecessary.

In principle, if fixed cost was zero, cumulative quantity would be sufficient to explain total unit cost and production rate would be irrelevant. But as fixed cost increases as a proportion of total cost, the impact of production rate should become important. This suggests that the relative bias of the LC and RA models may depend on the amount of fixed cost burden assigned to total cost.

Fixed cost burden was simulated by varying the percentage of total unit cost made up of fixed cost.[10] Three percentages were used in the simulation:

---

[9]See Conway and Schultz [9] for further elaboration of factors impacting learning rates.

[10]Operationally this is a bit complex, since both per unit variable and per unit fixed cost depend on other simulation inputs (cumulative quantity and production rate per period). The process of relating fixed cost to total

15%, 33%, and 50%. The different percentages can be viewed as simulating different degrees of operating leverage, of capital intensiveness, or of plant automation. The 15% level reflects the average fraction of price represented by fixed overhead in the aerospace industry, as estimated at one time by DoD [3].[11] The larger percentages are consistent with the trend toward increased automation [19].

iv) Production Rate Trend (PROTREND): When initiating a new product, it is not uncommon for the production rate per period to start low and trend upward to some "normal" level. This may be due both to the need to develop demand for the output or the desire to start with a small production volume, allowing slack for working bugs out of the production process. Alternatively, when a "new" product results from a relatively small modification of an existing product, sufficient customer demand or sufficient confidence in the production process may be assumed and full scale production may be initiated rapidly. In short, two different patterns in production volume may be exhibited early on when introducing a new item: a gradual growing trend toward full scale production or a level trend due to introduction at full scale production volume.

The simulation created two production trends during the model estimation period: "level" and "growth." These represented general trends (but as will become clear momentarily, variance around the general trend was introduced). The level trend simulated a production rate set at a "standard" 100% each period during model estimation. The growth trend simulated production rate climbing gradually to 100%. Details of the trends are in table 1.

v) Production Rate Instability/Variance (RATEVAR): Numerous factors, in addition to the general trend in output discussed above, may operate to cause period-to-period fluctuations in production rate. Manufacturers typically do not nave complete control over either demand for output or supply of inputs. Conditions in either market can cause instability in production rate. (Of

---

cost was as follows: First, a cumulative average per unit variable cost for all units produced during the estimation period was determined. Then a standard fixed cost per unit was set relative to the cumulative average per unit variable cost. For example, if standard fixed cost per unit was set equal to cumulative average variable cost per unit, then "on average" fixed cost would comprise 50% of total unit cost during the estimation period. Actual fixed cost per unit may differ from standard fixed cost per unit if the production rate (discussed later) was not at 100% of standard.

[11]In the absence of firm-specific cost data, the Cost Analysis Improvement Group in the Office of the Secretary of Defense treats 15% of the unit price of a defense system as representing fixed cost [22].

course, unstable demand, due to the uncertainties of annual budget negotiations, is claimed to be a major cause of cost growth during the acquisition of major weapon systems by the DoD). Production rate instability was simulated by adding random variance to each period's production rate during the estimation period. The amount of variance ranged from a coefficient of variation of .05 to .15 to .25. For example, if the production trend was level and the coefficient of variation was .05 then "actual" production rates simulated were generated by a normal distribution with mean equal to the standard production rate (100%) and sigma equal to 5%.

vi) Cost Noise/Variance (COSTVAR): From period to period there will be unsystematic, unanticipated, non-recurring, random factors that will impact unit cost. Changes in the cost, type or availability of input resources, temporary increases or decreases in efficiency, and unplanned changes in the production process are all possible causes. Conceptually such unsystematic factors can be thought of as adding random noise to unit cost. While unsystematic variation in cost cannot (by definition) be controlled, it is often possible to characterize different production processes in terms of the degree of unsystematic variation; some processes are simply less well-understood, more uncertain, and less stable than others.

Does bias depend on the stability of the process underlying cost? To investigate this question, random variance was added to the simulated costs generated from the cost function. The amount of variance ranged from a coefficient of variation of .05 to .15 to .25. For example, when the coefficient of variation was .25, then "actual" unit costs simulated were generated by a normal distribution with mean equal to cost from equation 9 and sigma equal to 25%.

vii) Future Production Level (FUTUPROD): Once a model is constructed (from data available during the estimation period), it is to be used to predict future cost. The production rate planned for the future may vary from past levels. Further growth may be planned. Cutbacks may be anticipated. Will the level of the future production rate affect the bias of the LC and RA models? Does one model tend to under (or over) estimate cost if cutbacks in production are anticipated and another if growth is planned? One might expect that inclusion of a rate term might be expected to reduce bias when production rate changes significantly (i. e., either growth or decline in the future period).

In the simulation, future production was set at three levels: low (50% of standard), same (100% of standard) and high (150% of standard). These simulate conditions of cutting back, maintaining or increasing production relative to the level of production existing at the end of the model estimation period.

    4. <u>The Cost Objective</u>: What is to be predicted? Up to this point the stated purpose of the study has been to evaluate bias when predicting future cost. But which future cost? Three alternatives were examined.

    i) Next period average unit cost: As the label suggests this is the average per unit cost of items in the production "lot" manufactured in the first period following the estimation period. Here the total cost of producing the output for the period is simply divided by the output volume to arrive at unit cost. Attention to this cost objective simulates the need to predict near term unit cost.

    ii) Total cost over a finite production horizon: The objective here is to predict the total cost of all units produced during a fixed length production horizon. Three periods was used as the length of the production horizon (one production lot produced each period). If the future production rate is low (high) then relatively few (many) units will be produced during the finite production horizon. Attention to this cost objective simulates the need to predict costs over some specific planning period, regardless of the volume to be produced during that planning period.

    iii) Total program cost: The objective here is to predict total cost for a specified number of units. If the future production rate is low (high) then relatively more (fewer) periods will be required to manufacture the desired output. The simulation was constructed such that at a low (same, high) level of future production six (three, two) future periods were required to produce the output. Attention to this cost objective simulates the need to predict total cost for a particular production program, regardless of the number of future periods necessary to complete the program.

    Examining each of these three cost objectives was deemed necessary to provide a well-rounded investigation of bias. However, the findings were the same across the three cost objectives. In the interest of space, the remainder of this paper will discuss the analysis and results only for the first cost objective, the average cost per unit for the next period's output.

    5. <u>The Measure of Bias</u>: A model specific measure of bias (BIAS) was determined separately for each (LC or RA) model as follows:

        BIAS = (PUC - AUC) ÷ AUC

where

        PUC =   Predicted unit cost from either the learning curve or the rate adjustment model.

        AUC =   Actual unit cost as generated by the cost function.

FIGURE 1. SIMULATION FLOWCHART

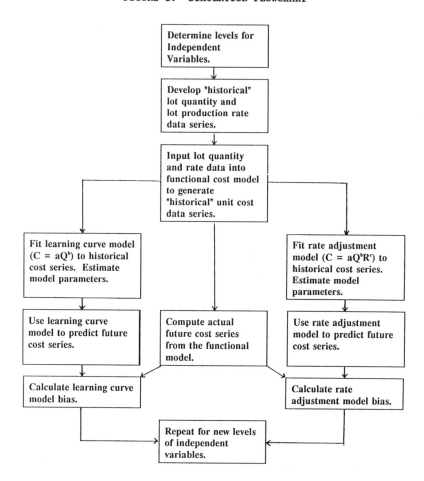

Positive values for BIAS indicate that a model overestimates actual future cost; negative values indicate underestimation. A model that is unbiased should, on average, produce values for BIAS of zero. BIAS represents the dependent variable in the statistical analysis. The research question then becomes: What factors or conditions explain variance in BIAS?

Figure 1 summarizes the complete simulation process leading up to the determination of BIAS. The simulation was run once for each possible combination of treatments. Given seven factors varied and three possible values for each factor (except for PROTREND which had two), there were 3 x 3 x 3 x 3 x 3 x 3 x 2 = 1458 combinations. Thus the simulation generated 1458 observations and 1458 values for BIAS for each of the two models.[12]

ANALYSIS AND FINDINGS

BIAS was evaluated using analysis of variance (ANOVA) to conduct tests of statistical significance. All main effects and first order (pairwise) interactions were examined. Findings with probability less than .01 were considered significant.

LC Model Bias. Table 2 provides ANOVA results addressing BIAS from the LC model. As shown, four main effects, DATAHIST, BURDEN, PROTREND, and FUTUPROD, are significant, indicating that values for BIAS are influenced by these treatment conditions. Table 3 summarizes BIAS values under the various conditions. Some interesting patterns are evident.

First, the overall mean BIAS across all observations is -.108. This means that, on average, the LC produces cost estimates that are about 11% too low.

Second, the mean BIAS for each treatment, for every variable of interest, is negative (with only one exception, when FUTUPROD is "high"). This means that the LC model bias toward underestimation is a consistent, pervasive phenomena. It is not driven by isolated conditions.

---

[12]In the simulation, just as in the real practice of cost analysis, it is possible for a model estimated on limited data to be very inaccurate, leading to extreme values for BIAS. If such outlier values were to be used in the subsequent analysis, findings would be driven by the outliers. Screening of the observations for outliers was necessary. During the simulation, if a model produced an BIAS value in excess of 100%, then that value was replaced with 100%. This truncation has the effect of reducing the impact of an outlier on the analysis while still retaining the observation as one that exhibited poor accuracy. Alternative approaches to the outlier problem included deletion instead of truncation and use of a 50% BIAS cutoff rather than the 100% cutoff. Findings were not sensitive to these alternatives.

TABLE 2

BIAS FROM LEARNING CURVE MODEL
ANALYSIS OF VARIANCE RESULTS

SOURCE	DF	SUM OF SQUARES	MEAN SQUARE	F VALUE
Model	85	56.92195	.6697	29.07
Error	1372	31.60432	.0230	PR>F:
Corrected Total	1457	88.52626		.0000

R^2		CV	BIAS MEAN	
.6430		140.35	−.1081	

SOURCE	DF	ANOVA SS	F VALUE	PR>F
DATAHIST	2	0.2937	6.38	0.0018*
VCRATE	2	0.0085	0.19	0.8311
BURDEN	2	0.3710	8.05	0.0003*
PROTREND	1	4.6998	204.03	0.0001*
RATEVAR	2	0.1167	2.53	0.0797
COSTVAR	2	0.0976	2.12	0.1205
FUTUPROD	2	47.0628	1021.54	0.0000*
DATAHIST*VCRATE	4	0.1184	1.29	0.2737
DATAHIST*BURDEN	4	0.0363	0.39	0.8124
DATAHIST*PROTREND	2	0.1280	2.78	0.0625
DATAHIST*RATEVAR	4	0.0265	0.29	0.8854
DATAHIST*COSTVAR	4	0.1503	1.63	0.1637
DATAHIST*FUTUPROD	4	0.1398	1.52	0.1944
VCRATE*BURDEN	4	0.0506	0.55	0.6990
VCRATE*PROTREND	2	0.0374	0.81	0.4435
VCRATE*RATEVAR	4	0.0623	0.68	0.6083
VCRATE*COSTVAR	4	0.1068	1.16	0.3271
VCRATE*FUTUPROD	4	0.2820	3.06	0.0159
BURDEN*PROTREND	2	0.3131	6.80	0.0012*
BURDEN*RATEVAR	4	0.0282	0.31	0.8738
BURDEN*COSTVAR	4	0.1631	1.77	0.1323
BURDEN*FUTUPROD	4	1.8751	20.35	0.0001*
PROTREND*RATEVAR	2	0.0176	0.38	0.6812
PROTREND*COSTVAR	2	0.0323	0.70	0.4955
PROTREND*FUTUPROD	2	0.3652	7.93	0.0004*
RATEVAR*COSTVAR	4	0.1570	1.70	0.1464
RATEVAR*FUTUPROD	4	0.0949	1.03	0.3902
COSTVAR*FUTUPROD	4	0.0855	0.93	0.4467

Third, in spite of the general tendency toward underestimation, the degree of bias does differ depending on the conditions. The effects of the

different conditions perhaps can be best demonstrated by a plot of BIAS
values by treatments. Figure 2 shows such a plot, with the (four signifi-

TABLE 3

LEARNING CURVE MODEL BIAS
BY MAIN EFFECTS

Independent Variable	BIAS for Each Level		
DATAHIST Value:	4	7	10
BIAS Mean:	−.096	−.100	−.128
VCRATE Value:	75%	85%	95%
BIAS Mean:	−.108	−.111	−.105
BURDEN Value:	15%	33%	50%
BIAS Mean:	−.086	−.119	−.120
PROTREND Value:	level	−	growth
BIAS Mean:	−.051	−	−.165
RATEVAR Value:	.05	.15	.25
BIAS Mean:	−.120	−.098	−.107
COSTVAR Value:	.05	.15	.25
BIAS Mean:	−.099	−.106	−.119
FUTUPROD Value	low	same	high
BIAS Mean:	−.344	−.070	.091

Overall Mean: −.108
Range of Group Means: −.344 to .091

cant) variables superimposed. In this plot, 1, 2, and 3 on the X-axis
reflect low, medium and high values for the independent variables (which are
taken from the left, middle and right columns of Table 3). Figure 2
reiterates the point made previously: BIAS is consistently negative (except
when FUTUPROD is high). More importantly, trends are evident:

a) Data History: Negative bias, the underestimation of future cost,
tends to increase as the number of observations available for estimating
model parameters (DATAHIST) increases. At first glance this seems counter-

FIGURE 2

PLOT OF LEARNING CURVE MODEL BIAS
BY MAIN EFFECTS

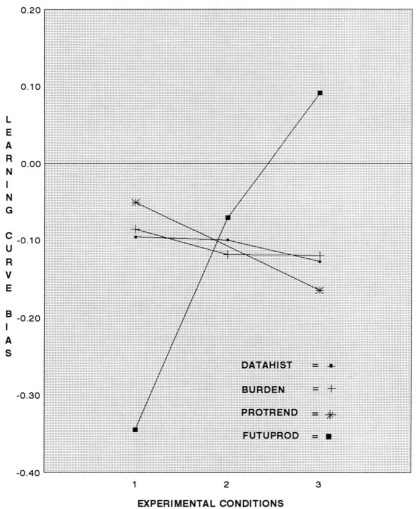

intuitive. Traditional wisdom says that having more data available leads to better parameter estimates and better model forecasts. But that is true only if a model is correctly specified. This issue will be discussed further later.

b) Fixed Cost Burden: Negative bias tends to increase as the proportion of total costs comprised of fixed costs (BURDEN) increases. This result is perhaps not surprising. In the underlying cost phenomena being modeled, learning impacts the incurrence of variable costs, not fixed costs. It is plausible that the LC model would become more biased as fixed costs increase.

c) Past Production Trend: The negative bias is considerably stronger if the rate of production was growing, rather than level during the model estimation period. This is not difficult to explain. An increasing production rate during the model estimation period will result in a steadily declining fixed cost per unit. An LC model will interpret this rate effect as a learning effect, and overestimate the degree of learning actually occurring. Future forecasts of cost will thus be biased downward.

d) Future Production Level: As the production rate, during the period for which costs are being forecast, shifts from "low" to "high", the LC model shifts from strongly underestimating to overestimating cost. In short, there is an inverse relationship between future production level and the bias toward underestimation. This effect is to be expected. Higher (lower) future production will result in lower (higher) fixed cost, and total cost, per unit, creating a tendency toward positive (negative) bias for any cost estimate.

Note that the only time cost is overestimated by the LC model is when future production level is high. The LC is still biased toward underestimation, but if the future production level increases enough to reduce per unit fixed cost enough, the tendency toward underestimation is masked by the offsetting tendency toward reduced actual per unit cost.

In addition to these main effects, the Table 2 ANOVA results indicated that pairwise interactions involving BURDEN, PROTREND and FUTUPROD are also significant; not only does BIAS depend on these three variables, it depends on how they interact. These interactions are illustrated in Figures 3, 4 and 5.

Figure 3, the interaction between Fixed Cost Burden and Production Rate Trend, merely reinforces previous findings: Negative bias tends to be greater when burden is higher or when the production rate grows during the model estimation period. The figure just indicates that the combination of

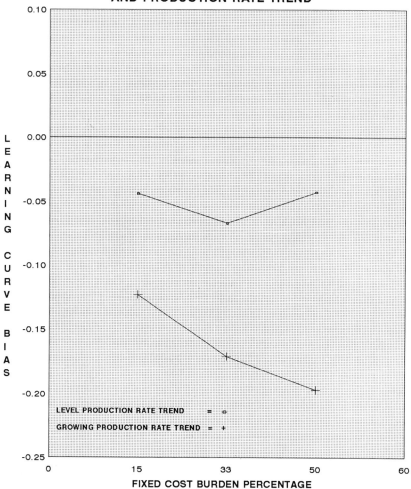

FIGURE 3
LEARNING CURVE MODEL BIAS

**INTERACTION OF FIXED COST BURDEN
AND PRODUCTION RATE TREND**

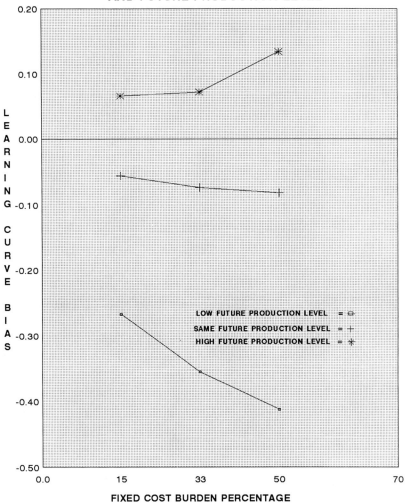

FIGURE 4
LEARNING CURVE MODEL BIAS

INTERACTION OF FIXED COST BURDEN
AND FUTURE PRODUCTION LEVEL

FIGURE 5

LEARNING CURVE MODEL BIAS

**INTERACTION OF PRODUCTION RATE TREND
AND FUTURE PRODUCTION LEVEL**

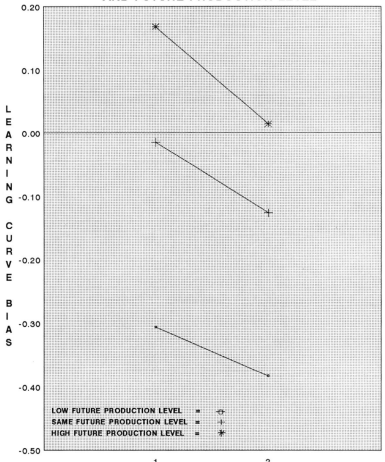

LEVEL ← PAST PRODUCTION TREND → GROWTH

these two conditions--high burden coupled with growing production volume--
magnifies the negative bias.

Figure 4, the interaction between Fixed Cost Burden and Future Production
Level, clearly reinforces the previously noted inverse relationship between
future production level and the bias toward underestimation. But findings
concerning Burden now appear conditional. High burden increases the tendency
toward underestimation, if future production level is low. But high burden
increases the tendency toward overestimation when future production level is
high. In short, increasing fixed cost burden magnifies the biasing effect--
in either direction--caused by shifts in the future production level.

Figure 5 shows the interaction between the production trend during the
model estimation period and the future production level during the forecast
period. The most interesting observation concerns the two points where BIAS
is close to zero. These occur when a) a "level" production trend is coupled
with the "same" level in the future forecast period, and b) a "growing"
production trend is coupled with a "high" level in the forecast period.
Consistency characterizes both situations; the production rate is either
consistently level or consistently increasing throughout the joint estima-
tion/forecast periods. In contrast, the greatest bias occurs when a
"growing" production trend is coupled with a "low" level in the future
forecast period. Here an inconsistent pattern, a shift from increasing to
decreasing production rate, causes severe underestimation of cost.

RA Model Bias. Table 4 provides ANOVA results addressing BIAS from the
RA model. Table 5 summarizes BIAS values under the various experiment
conditions. Two findings are evident. First the overall mean BIAS for all
observations is only -.0016. Thus, on average, the RA model exhibits no
bias. Second, this absence of bias is evident for all treatments across all
variables of interest. There are no significant main effects in the ANOVA
results and group means for BIAS in table 5 range only from -.021 to .026.
Thus the overall absence of bias is not caused by positive bias under some
conditions offsetting negative bias under other conditions. Rather the
absence of noticeable bias exists across the various treatments.

There is one statistically significant first order interaction in the
ANOVA. Figure 6 plots this interaction between Production Rate Trend and
Future Production Level. Two points seem noteworthy. First, the greatest
bias occurs when a "growing" production trend during the model estimation
period is coupled with a "low" production level in the forecast period. So,
as with the LC model, a shift from increasing to decreasing production causes

bias to occur. Second, in spite of this interaction result being statistically significant, the magnitude of bias evident is far less than with the LC

TABLE 4

BIAS FROM RATE ADJUSTMENT MODEL
ANALYSIS OF VARIANCE RESULTS

SOURCE	DF	SUM OF SQUARES	MEAN SQUARE	F VALUE
Model	85	11.18626	.1316	1.08
Error	1372	166.9451	.1217	PR>F:
Corrected Total	1457	178.1314		.2919

R^2		CV		BIAS MEAN
.0628		21638.82		−.0016

SOURCE	DF	ANOVA SS	F VALUE	PR>F
DATAHIST	2	0.1779	0.73	0.4815
VCRATE	2	0.3435	1.41	0.2441
BURDEN	2	0.0539	0.22	0.8012
PROTREND	1	0.2335	1.92	0.1662
RATEVAR	2	0.2986	1.23	0.2934
COSTVAR	2	0.5567	2.29	0.1019
FUTUPROD	2	0.3965	1.63	0.1964
DATAHIST*VCRATE	4	0.3066	0.63	0.6412
DATAHIST*BURDEN	4	0.0866	0.18	0.9498
DATAHIST*PROTREND	2	0.0972	0.40	0.6706
DATAHIST*RATEVAR	4	0.3802	0.78	0.5373
DATAHIST*COSTVAR	4	0.0617	0.13	0.9727
DATAHIST*FUTUPROD	4	0.2723	0.56	0.6921
VCRATE*BURDEN	4	0.6156	1.26	0.2818
VCRATE*PROTREND	2	0.1873	0.77	0.4633
VCRATE*RATEVAR	4	0.3605	0.74	0.5642
VCRATE*COSTVAR	4	0.1389	0.29	0.8875
VCRATE*FUTUPROD	4	1.3745	2.82	0.0238
BURDEN*PROTREND	2	0.0470	0.19	0.8243
BURDEN*RATEVAR	4	0.3449	0.71	0.5860
BURDEN*COSTVAR	4	0.3527	0.72	0.5751
BURDEN*FUTUPROD	4	0.6125	1.26	0.2844
PROTREND*RATEVAR	2	0.1738	0.71	0.4897
PROTREND*COSTVAR	2	0.2152	0.88	0.4132
PROTREND*FUTUPROD	2	1.1777	4.84	0.0080*
RATEVAR*COSTVAR	4	0.1900	0.39	0.8156
RATEVAR*FUTUPROD	4	1.5652	3.22	0.0122
COSTVAR*FUTUPROD	4	0.5640	1.16	0.3273

model. In a comparative sense, the RA model still does not appear to create
a bias problem.

TABLE 5

RATE ADJUSTMENT MODEL BIAS
BY MAIN EFFECTS

Independent Variable	BIAS		
DATAHIST Value:	4	7	10
BIAS Mean:	.004	.008	-.017
VCRATE Value:	5%	85%	95%
BIAS Mean:	-.021	-.000	.016
BURDEN Value:	15%	33%	50%
BIAS Mean:	-.004	-.008	.007
PROTREND Value:	level	-	growth
BIAS Mean:	-.014	-	.011
RATEVAR Value:	.05	.15	.25
BIAS Mean:	.016	-.002	-.019
COSTVAR Value:	.05	.15	.25
BIAS Mean:	-.019	-.011	.026
FUTUPROD Value	low	same	high
BIAS Mean:	.015	.004	-.024

Overall Mean: -.0016
Range of Group Means: -.021 to .026

Additional Analysis of LC Bias: The findings that the degree of bias in
the LC model is dependent on PROTREND and FUTUPROD is not completely
surprising. Both variables reflect how production rate varies from period to
period, and the LC model does not include a rate term.[13]

---

[13]This does not mean the finding is without interest. Many researchers
and cost analysts [e.g., 12] have noticed that empirically there is often
high colinearity between cumulative quantity and production rate. This

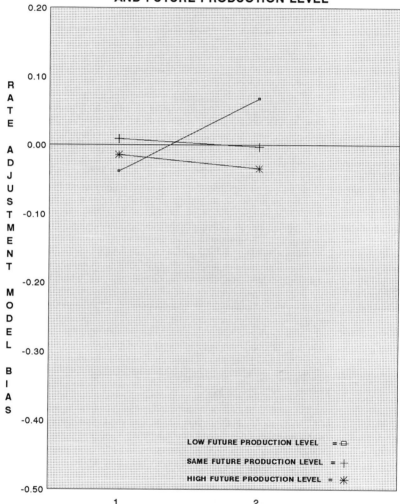

FIGURE 6
RATE ADJUSTMENT MODEL BIAS
INTERACTION PRODUCTION RATE TREND
AND FUTURE PRODUCTION LEVEL

The findings that LC model bias also depends on DATAHIST and BURDEN merit a bit more attention. To further investigate, some addition simulations were run under "ideal" conditions, where impacts on cost caused by the other variables were suppressed. More specifically equation 9 was used to generate cost series where a) production rate was level during the model estimation period, b) production rate stayed at the same level during the cost forecast period, c) random noise in cost was set at zero, and d) production rate variance was set at zero. Only VCRATE, BURDEN and DATAHIST were varied. Again LC models were fit to the cost series and then estimated future costs were compared with actual future costs.

i) The Concave Curve: Figure 7 shows a log-log plot of residuals (actual minus estimate cost) by quantity for one illustrative situation (where VCRATE = 75%; BURDEN = 50%; DATAHIST = 7). Recall that a central assumption of a learning curve is that cost and quantity are log linear. Figure 7 shows cost as estimated and predicted by the LC model as a horizontal line (abscissa of zero), while the plot of the residuals displays the pattern of actual costs relative to the LC line. Note that actual costs are not log linear with quantity; instead an obvious concave curve is evident. This pattern is not a result of the particular values for VCRATE, BURDEN, and DATAHIST; the same pattern was evident for all other combinations of variable values examined.

The vertical line in the figure separates the seven cost observations used to estimate the LC model, on the left, from three future costs the model is used to predict, on the right. The concavity of the actual cost curve results in each successive actual cost diverging increasingly from the LC model prediction. The conclusion to be drawn is that whenever a learning curve is used to model a cost series that includes some fixed cost component (some component that is not subject to learning), then a log linear model is being fit to a log concave phenomena. A systematic bias toward underestimation of future cost is inherent in the LC model.

ii) Bias Patterns: Table 6 lists measures of BIAS for various combinations of BURDEN and VCRATE. The absolute magnitude of the BIAS values is not important; three patterns in the table are. First, reading $BIAS_1$ through $BIAS_4$ values across any row reiterates the pattern exhibited in figure

---

colinearity has been argued to make production rate a somewhat redundant variable in a model, leading to unreliable parameter estimates when the model is estimated and providing little incremental benefit when the model is used for forecasting future cost. The current findings suggest that one role of a production rate variable in a model is to reduce model bias.

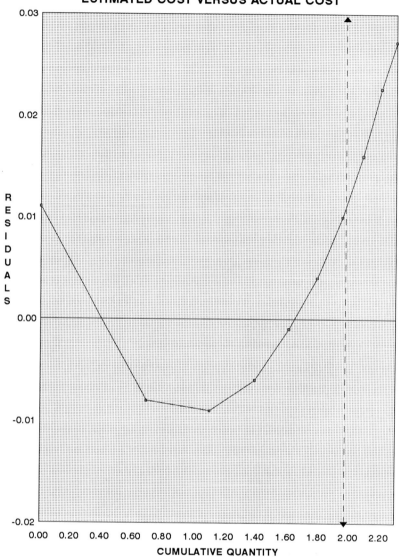

FIGURE 7
ESTIMATED COST VERSUS ACTUAL COST

4. Bias increases when estimating each additional future unit. This suggests that the further into the future the LC model is used to estimate costs, the greater the underestimation will be.

TABLE 6

BIAS PATTERNS FROM THE LC MODEL
(At selected values for BURDEN and VCRATE)

VCRATE	BURDEN	$BIAS_1$	$BIAS_2$	$BIAS_3$	$BIAS_4$
75%	10%	-.00544	-.00758	-.00973	-.01186
	20%	-.00973	-.01348	-.01720	-.02087
	30%	-.01287	-.01772	-.02250	-.02719
	40%	-.01487	-.02037	-.02575	-.03099
	50%	-.01570	-.02138	-.02692	-.03229
	60%	-.01531	-.02076	-.02604	-.03113
	70%	-.01365	-.01843	-.02303	-.02747
	80%	-.01063	-.01429	-.01781	-.02119
85%	10%	-.00176	-.00243	-.00310	-.00376
	20%	-.00315	-.00435	-.00552	-.00668
	30%	-.00418	-.00573	-.00727	-.00877
	40%	-.00482	-.00660	-.00834	-.01005
	50%	-.00507	-.00692	-.00873	-.01050
	60%	-.00492	-.00670	-.00843	-.01012
	70%	-.00436	-.00592	-.00743	-.00891
	80%	-.00336	-.00456	-.00571	-.00683
95%	10%	-.00018	-.00025	-.00032	-.00038
	20%	-.00033	-.00045	-.00056	-.00068
	30%	-.00043	-.00058	-.00074	-.00089
	40%	-.00049	-.00067	-.00084	-.00101
	50%	-.00051	-.00069	-.00088	-.00105
	60%	-.00049	-.00067	-.00084	-.00101
	70%	-.00043	-.00058	-.00073	-.00088
	80%	-.00032	-.00044	-.00056	-.00067

NOTE: DATAHIST = 7. $BIAS_1$ is the bias in forecasting the cost of the first unit produced after the model estimation period; $BIAS_2$ relates to the second unit, etc.

Second, moving from the bottom, to the middle, to the top panel of the table--from VCRATE 95%, to 85%, to 75%--it is clear that BIAS increases. The general pattern suggested is that as the "true" underlying learning rate (of the portion of total cost subject to learning) increases, the tendency of the LC model to underestimate future cost also increases.

Third, read down any column to observe the pattern of BIAS values as BURDEN increases from 10% to 80% of total cost. Negative bias consistently increases with increases in fixed cost burden--up to a point--then negative bias decreases with further increases in burden. The turn around point for all observations is when burden is 50%. This confirms the finding from the earlier ANOVA test, that bias increases with burden, but indicates that that pattern holds only when fixed cost is less than half of total cost; the pattern is not universal. This reversal is perhaps understandable. Consider the two extremes. If BURDEN = 0%, then all cost would be variable, all cost would be subject to learning, an LC model would be a correct specification of the "true" underlying cost function, and zero bias would result. If BURDEN = 100% then all cost would be fixed, no cost would be subject to learning, an LC model would again be a correct specification of the "true" underlying cost function (which would be a learning curve with slope of zero--no learning), and zero bias would result. Only when costs--some subject to learning, some not--are combined does the bias result. And the bias is at a maximum when the mixture is about fifty-fifty.

iii) Bias and Estimated LC Slope: Recall that the total cost of any unit produced depends on both VCRATE, which determines the learning experienced by the variable cost portion of total cost, and BURDEN, which determines the magnitude of the fixed cost portion of total cost. Given the findings that BIAS depends on both VCRATE and BURDEN raises an interesting practical question. In many circumstances, cost analysts may not have access to detailed cost data and hence may not "know" the values for VCRATE and BURDEN in a real world cost problem being analyzed. In fact, the point of fitting a learning curve to cost data is typically to arrive at a summary description of an unknown cost function. What is observable by the analyst is an _estimated_ learning curve slope for a given observable total cost series. Is there a relationship between estimated LC slope and BIAS? The nature of that relationship is not obvious. Ceteris paribus, as VCRATE become steeper, estimated LC slope will become steeper as well. Given the tendency of BIAS to vary with VCRATE, this suggests that BIAS will increase as estimated LC gets steeper. But, ceteris paribus, as BURDEN increases, estimated LC slope will become more shallow. Given the tendency of BIAS to first increase, then decrease with increases in BURDEN, the relationship between estimated LC slope and BIAS is ambiguous.

Figure 8 plots BIAS against estimated LC slope (generated for combinations of VCRATE, varied from 70% to 95%; BURDEN, varied from 10% to 80%; DATAHIST = 7). Note that the scatter diagram is not tightly clustered along

FIGURE 8
PLOT OF BIAS VERSUS
ESTIMATED LC SLOPE

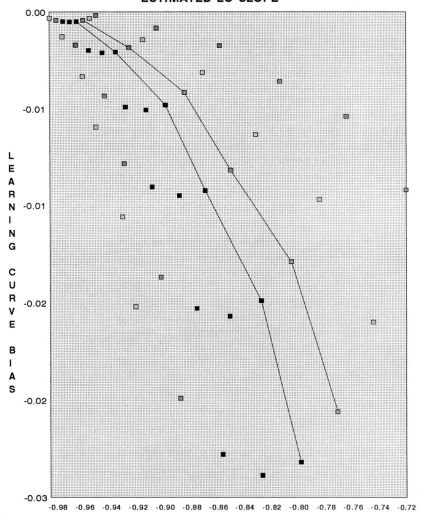

SLOPE

any trend line. In the most general sense, there is no strong relationship between estimated LC slope and bias. But consider the segment of the plot falling within the boundaries formed by the two dotted lines. These represent the boundaries for BIAS when BURDEN is constrained, in this case, to fall between 30-40%. Given that burden is assumed to vary through only a small range, then there is a strong empirical relationship: steeper estimated LC slopes are associated with a greater tendency toward underestimation of cost.

iv) Bias and Data History: Table 7 explores the impact of DATAHIST on BIAS. Here BIAS is measured for cost forecasts from models estimated on n data points, where n is varied from 4 through 10. For each model, BIAS is measured for n + 1, n + 2, etc. Recall from the earlier ANOVA results that bias increased as DATAHIST increased. This lead to the somewhat counterintuitive conclusion that LC models get progressively more biased the more observations there are available on which to fit the model. Two patterns in the table confirm this finding but clarify its implications.

TABLE 7

THE IMPACT OF DATAHIST ON BIAS

DATAHIST	$BIAS_5$	$BIAS_6$	$BIAS_7$	$BIAS_8$	$BIAS_9$	$BIAS_{10}$	$BIAS_{11}$
4	-.0125	-.0204	-.0281	-.0355	-.0425	-.0493	-.0557
5		-.0137	-.0207	-.0275	-.0341	-.0404	-.0464
6			-.0148	-.0211	-.0271	-.0330	-.0387
7				-.0157	-.0214	-.0269	-.0323
8					-.0165	-.0217	-.0268
9						-.0172	-.0220
10							-.0178

NOTE:
VCRATE =   75%
BURDEN =   50%
$BIAS_n$: $BIAS_n$ measures the bias associated with estimating the cost of the nth unit in the cost series.

First, observe the BIAS values in the diagonal (top left to bottom right) of the table. BIAS consistently increases. The prediction of, say, the 7th cost in a series using an LC model estimated on the first six costs will be more biased than the prediction of the 6th cost using a model estimated on the first five. Bias in predicting the n + 1 cost does increase with n. (This is the same finding as from the ANOVA.)

But observe also the BIAS values in any column. BIAS consistently decreases as DATAHIST increases. The prediction of, say, the 7th cost using a model estimated on the first six costs is less biased than the prediction of that same 7th cost using a model estimated on only the first five. In short, given a task of forecasting a _specific_ given cost, ceteris paribus, it is always beneficial to use as many data points as are available to estimate the LC model.

SUMMARY AND CONCLUSIONS

The central purpose of this study was to examine bias in estimating future cost from two models commonly used in cost estimation. The analysis simulated prediction for both the traditional learning curve and a rate adjustment model, and evaluated bias under varying conditions. The broadest finding was that the rate adjustment model provided cost estimates that were unbiased, while the learning curve model consistently produced estimates that understated actual cost. Most additional findings concerned the conditions related to bias in the learning curve model:

- The cause of the bias is the existence of fixed cost in total cost. The learning curve assumes a log linear relationship between cost and quantity, which does not hold when fixed cost (not subject to learning) is present.

- The bias increases as the proportion of fixed cost in total cost increases--up to the point where fixed cost comprises about 50% of total cost--after that further increases in fixed cost reduce bias. This finding would appear to be relevant given the trend in modern production processes toward increasing automation and hence an increasing fixed component in total cost.

- The degree of bias is affected by the production rate during both the period of model estimation and the period for which costs are forecast. A consistent production rate trend throughout these periods minimizes bias. A shift in production rate trend, particularly to a cutback in volume, magnifies bias. This finding would appear to be relevant to cost estimators analyzing programs where cutbacks are anticipated.

- Assuming the proportional relationship between fixed and variable components of total cost does not vary greatly, bias is greater when the estimated learning curve slope is steeper.

- The bias problem is not diminished as more observations become available to estimate the learning curve. In fact the degree of bias increases as the number of observations increases.

— The degree of bias increases the further into the future predictions are made. Next period cost is somewhat underestimated; cost two periods in the future is underestimated to a greater degree, etc.

Some of the conclusions are a bit ironic. One typically expects to improve forecasting when more data is available for model estimation. The findings here suggest that bias grows worse. One typically expects future costs to decline most rapidly when past costs have exhibited a high rate of learning. The findings here suggests that such circumstances are the ones most likely to result in actual costs higher than forecasted.

Caution should be exercised in drawing direct practice-related implications from these findings. The finding that the rate adjustment model is unbiased while the learning curve is biased does not mean that the rate model should always be preferred to the learning curve model. Bias is only one criteria for evaluating a cost estimation model. Consider accuracy. Evidence indicates that under some circumstances learning curves are more accurate than rate adjustment models [21]. Thus model selection decisions would need to consider (at a minimum) tradeoffs between bias and accuracy. An accurate model with a known bias, which could be adjusted for, would typically be preferable to an inaccurate, unbiased model.

The conclusions of any study must be tempered by any limitations. The most prominent limitation of this study is the use of simulated data. Use of the simulation methodology was justified by the need to create a wide range of treatments and maintain control over extraneous influences. This limitation suggests some directions for future research.

-- Re-analyze the research question while altering aspects of the simulation methodology. For example, are findings sensitive to the cost function assumed?

-- Address the same research question using actual cost and production rate data. Are the same findings evident when using "real-world" data?

Providing confirmation of the findings by tests using alternative approaches would be beneficial.

Additional future research may be directed toward new, but related, research questions.

-- Investigate other competing models or approaches to cost prediction. Perhaps bias can be reduced by using some version of a "moving average" prediction model. Can such a model outperform both the learning curve and the rate adjustment approach? If so, under what circumstances?

-- Investigate tradeoffs between various characteristics of cost estimation models, such as bias versus accuracy.

REFERENCES

1. Alchian, A. (1963), "Reliability of Progress Curves in Airframe Production," _Econometrica_, Vol. 31, pp. 679-693.

2. Asher, H. (1956), _Cost-Quantity Relationships in the Airframe Industry_, R-291, RAND Corporation, Santa Monica, CA.

3. Balut, S. (1981), "Redistributing Fixed Overhead Costs," _Concepts_, Vol. 4, No. 2, pp. 63-72.

4. Balut, S., T. Gulledge, Jr., and N. Womer (1989), "A Method of Repricing Aircraft Procurement," _Operations Research_, Vol. 37, pp. 255-265.

5. Bemis, J. (1981), "A Model for Examining the Cost Implications of Production Rate," _Concepts_, Vol. 4, No. 2, pp. 84-94.

6. Boger, D. and S. Liao (1990), "The Effects of Different Production Rate Measures and Cost Structures on Rate Adjustment Models," in W. Greer and D. Nussbaum, editors, _Cost Analysis and Estimating Tools and Techniques_, Springer-Verlag, New York, 1990, pp. 82-98.

7. Cheney, W. (1977), _Strategic Implications of the Experience Curve Effect for Avionics Acquisition by the Department of Defense_, Ph. D. Dissertation, Purdue University, West Lafayette, IN.

8. Cochran, E. (1960), "New Concepts of the Learning Curve," _Journal of Industrial Engineering_," Vol. 11, 317-327.

9. Conway, R. and A. Schultz (1959), "The Manufacturing Progress Function," _Journal of Industrial Engineering_, 10, pp. 39-53.

10. Cox, L. and J. Gansler (1981), "Evaluating the Impact of Quantity, Rate, and Competition," _Concepts_, Vol. 4, No. 4, pp. 29-53.

11. Dorsett, J. (1990), "The Impacts of Production Rate on Weapon System Cost," paper presented at the Joint Institute of Cost Analysis/National Estimating Society National Conference, Los Angeles, CA, June 20-22.

12. Gulledge, T. and N. Womer (1986), _The Economics of Made-to-Order Production_, Springer-Verlag, New York, NY.

13. Hirsch, W. (1952), "Manufacturing Progress Functions," _The Review of Economics and Statistics_, Vol. 34, pp. 143-155.

14. Large, J., H. Campbell and D. Cates (1976), _Parametric Equations for Estimating Aircraft Airframe Costs_, R-1693-1-PA&E, RAND Corporation, Santa Monica, CA.

15. Large, J., K. Hoffmayer, and F. Kontrovich (1974), _Production Rate and Production Cost_, R-1609-PA&E, The RAND Corporation, Santa Monica, CA.

16. Levenson, G., et. al. (1971), _Cost Estimating Relationships for Aircraft Airframes_, R-761-PR, RAND Corporation, Santa Monica, CA.

17. Liao, S. (1988), "The Learning Curve: Wright's Model vs. Crawford's Model," _Issues in Accounting Education_, Vol. 3, No. 2, pp. 302-315.

18. Linder, K. and C. Wilbourn (1973), "The Effect of Production Rate on Recurring Missile Costs: A Theoretical Model," _Proceedings_, Eighth Annual Department of Defense Cost Research Symposium, Airlie VA, compiled by Office of the Comptroller of the Navy, 276-300.

19. McCullough, J. and S. Balut (1986), _Defense Contractor Indirect Costs: Trends, 1973-1982_, IDA P-1909, Institute for Defense Analysis, Alexandria, VA.

20. Moses, O. (1990a), _Extensions to the Learning Curve: An Analysis of Factors Influencing Unit Cost of Weapon Systems_, Naval Postgraduate School Technical Report, NPS-54-90-016, Monterey, CA.

21. Moses, O. (1990b), "Learning Curve and Rate Adjustment Models: Comparative Prediction Accuracy under Varying Conditions," in R. Kankey and J. Robbins, eds., _Cost Analysis and Estimating: Shifting U.S. Priorities_, Springer Verlag, 1990, pp. 65-102.

22. Pilling, D. (1989), _Competition in Defense Procurement_, The Brookings Institution, Washington DC, p. 35.

23. Smith, C. (1980), _Production Rate and Weapon System Cost: Research Review, Case Studies, and Planning Model_, APR080-05, U. S. Army Logistics Management Center, Fort Lee, VA.

24. Smith, C. (1981), "Effect of Production Rate on Weapon System Cost," _Concepts_, Vol. 4, No. 2, pp. 77-83.

25. Smith, L. (1976), _An Investigation of Changes in Direct Labor Requirements Resulting From Changes in Airframe Production Rate_, Ph. D. dissertation, University of Oregon, Eugene, OR.

26. Smunt, T. (1986), "A Comparison of Learning Curve Analysis and Moving Average Ratio Analysis for Detailed Operational Planning," _Decision Sciences_, Vol. 17, No. 4, Fall, pp. 475-494.

27. Washburn, A. (1972), "The Effects of Discounting Profits in the Presence of Learning in the Optimization of Production Rates," _AIIE Transactions_, 4, pp. 255-313.

28. Wetherill, G. (1986), _Regression Analysis with Applications_, Chapman and Hall, New York.

29. Womer, N. (1979), "Learning Curves, Production Rate and Program Costs," _Management Science_, Vol. 25, No. 4, April, pp. 312-319.

30. Wright, T. (1936), "Factors Affecting the Cost of Airplanes," _Journal of Aeronautical Sciences_, Vol. 3, pp. 122-128.

31. Yelle, L. (1979), "The Learning Curve: Historical Review and Comprehensive Survey," _Decisions Sciences_, Vol. 10, No. 2, April, pp. 302-328.

The Production Rate Impact Model: IRATE

Schuyler C. Lawrence
EER Systems

ABSTRACT

The IRATE model is an analytic procedure for measuring the direct impact of production rate upon a cost improvement curve (CIC) slope. The IRATE process involves a steepening of the assembly's CIC slope, which is experienced in parallel with or after the installation of all non-recurring items, (e.g., new setups, tool investments, facilities, process changes), specifically added for the higher production rate program. The IRATE effect abstracts the progressive characteristic of recurring events, bringing together the horizontal and vertical "learning" synergisms of production lines.

INTRODUCTION

IRATE is a process working within the standard learning curve process. Because of its block-to-block dependency IRATE complements the Wright learning curve methodology. It is this dependency which not only smooths slope variation, but also integrates the collective know-how or "memory" experienced over the changing production span.

Like the standard learning curve theories, IRATE deals with measuring learning curve slopes. Unlike standard learning curve theory, IRATE is a non-logarithmic solution that measures learning curve percentage as a function of production rate change. Slopes are treated as differentials with each successively added production block, providing the production rate changes. Therefore, it is a means-to-an-end in the learning curve process. As a linear transformation, the model has parametric simplicity and can be exercised in two ways:

1.   IRATE, as a regression transformation technique, "fits" empirical data. Unlike standard learning regressions, logarithmic transformations are not necessary.

2.   IRATE solutions can be constructed from a single point application. Program CIC assumptions can be initiated at the "no learning", 100% slope reference, and calibrated by production rate advances.

IRATE is never a substitute for the standard learning curve process, but a step in that process: IRATE fits slope changes as a function of production rate changes, suggesting multiple-slope linkages. It is the tangent-to-tangent progression that describes a total CIC performance, which seldom if ever results as a straight line on the log-log grid.

THE IRATE MODEL DEFINED

The linear model, $Y = a + b(X)$, being the simplest mathematical form, is a condition for the investigation. The balance of the IRATE Model derivation is a description of a transform process defined as a function of (X).

The CIC slope (S) is an exponent relating to a constant percentage of improvement acquired with the doubling of quantity,

$$S = \ln(LCF/100) / \ln 2 \tag{1}$$

where the percentage or learning curve function (LCF) is equated to the linear condition.

$$LCF = a + b(X) \tag{2}$$

The (X) term is a function of the ratio of the initial rate (C) to the sum of the initial rate and the succeeding rate (R).

$$(X) = C / (C + R) \tag{3}$$

The transform is also comfortably realized as the reciprocal of the succeeding rate to the initial rate plus one, which simply converts the relationship to a hyperbola.

$$(1 / X) = [(R / C) + 1]^{-1} \tag{4}$$

The added constant of one to the ratio of rates constrains the intercept and localizes applications to the upper right quadrant.

The IRATE model calculates the learning curve (percentage) function (LCF) for an accumulation of runs or lots which tend toward an asymptotic LCF (%) limit at a high production rate. The ratio of nominally-expected production rates for all runs to the starting production rate is measured for each calculation of the equation:

$$LCF_n = a + b[C / (C + R_n)] \tag{5}$$

IRATE MODEL CHARACTERISTICS

When C = R, (X) = 0.5, at all times. In the algebraic balance of parameters, where the LCF does not exceed 100%, this means that parameter "b" varies twice as much as "a" for every change to "b". The condition of "no learning" occurs when b = 0, and a = 100%. If C > R, as in the case of spares production, LCF may exceed 100%, displaying "toe-up" properties. However, for the most practical applications, C is equal to, or less than R. Finally, when R = 0, (b + a) = 100%.

IRATE MODEL PARAMETERS

With sufficient ground-work layed, a discussion of parameter sensitivities is introduced prior to illustrating practical applications. The fixed parameters of the IRATE model are the coefficient (a) and (b). The variables are C and R. C is the initializer and is generally fixed for subsequent calculations. This makes it at least a circumstantial parameter. The asymptotic parameter (a), when it is not an output from a regression analysis, can be handled as a function that states a position about quantifying collective efficiency measures. These efficiency measures, however, are a method bringing together lines and stations interaction, at a lower level of evaluation (see detailed discussion later in text). The (a) parameter is essential to the successful IRATE solution, however it is a parameter that falls into a very precise percentage range, and therefore not as sensitive as other parameters.

The (b) parameter which provides the margin of variability due to rate change, over the asymptote, is usually controllable when the regression fit includes data pairs [LCF = 100; C/(C+R) = 0.5]. This also presumes a fit with significant correllation, which cannot realistically be the case, all the time. Therefore, when the fit is good, one of the conditions for a satisfactory (b) parameter is that one-half its value added to (a), is approximately 100%.

The R variable is always the production rate (per year) that is compared to the rate initializer, C. This comparison or ratio, is usually made for production lots in succession, however, it need not be: the important thing to remember is, that whatever the comparison of C and R, the LCF output is a cumulative average statement for a CIC segment contingent upon a progressive build-up of production capabilities. This is essential information when fitting the model in a regression analysis. Advanced programs estimating should assume the use of R as an average program rate, compared to its "kick-off" or Full Scale Development/Pilot Production rate.

IRATE MODEL APPLICATIONS

Three ways of applying the IRATE Model will be presented. The first of these demonstrates the simple application of IRATE to an advanced program cost estimate. The second shows a way of pre-supposing the parameters of a regression fit, using only the first production lot data. The third application adds the data excluded for the 2nd application, with a complete regression analysis for formulating the parameters.

CASE #1: Evaluating The Procurement Rate Cut

Assume that a program has undergone a critical revision to its procurement, requiring a cut in yearly production rate from 50 per year to 10 per year. Its initial production rate, C, was 2 per year, during its Full Scale Development/Pilot Production Transition. Evaluations of comparable programs show the CIC potential at greater than 50 per year rate to be about 85% at the optimum rate. With this information, determine the CIC slope for the lower, 10 per year production program:

(1) Determine the (b) parameter in the range between no learning (100%) at two per year, and optimal rate (85%) at > 50 per year rate.

$$b = (100\% - 85\%) / [2 / (2 + 2)] = 15 / 0.5 = 30$$

(2) Solve the high production rate option at 50 per year.
$$\{[2 / (2 + 50)] * 30\} + 85 = 86.154\%$$

(3) Solve the reduced rate production option at 10 per year.
$$\{[2 / (2 + 10)] * 30\} + 85 = 90\%$$

The lower-rate production program results in a flatter CIC performance.

CASE #2: The Single Block IRATE Solution

There is a simple procedure for generating an IRATE equation from only one production block, given a few empirical observations and some maneuvering of the model's parametric characteristics. This shall be demonstrated by some actual data provided by [3]. The Patriot Missile Program has been selected for the purpose. Specifically, the first of five successive procurement blocks shall provide the information necessary for the illustration, the first block consisting of three years of rates averaging 141 units per year. A 96.742% cumulative average slope is calculated through this three-year run. Since we know the LCF for the first run, and the initial average rate (C), the balance of fixed parameters can be hypothesized as follows:

(1) The (b) parameter is (theoretically) a constant, if the analyst accepts the practical program CIC range to be between 100% (no improvement) and the Stanford B Curve [2] slope limit of -0.5, or 70.7%. If this practical limitation on program learning is accepted, then calculating (b) assumes a stagnant, no-follow-on condition,

$$[141 / (141 + 0)] * b + 70.7 = 100$$

$$b = 29.3 \qquad \text{(CIC possibilities range)}$$

(2) Using the hypothetical (b) coefficient, the first block CIC slope of 96.742%, and an assumed no-rate-change follow-on,

$$[141 / (141 + 141)] * 29.3 + a = 96.742$$

$$a = 82.092 \text{ (CIC asymptote adjustment)}$$

(3) The theoretical IRATE model for the Patriot Program can then be constructed, per equation (5),

$$\{[C / (C + R_n)] * 29.3\} + 82.092 = LCF_n$$

CASE #3: The Regression Analysis Fit

When the data analysis initiates from the first block, the IRATE fit becomes more valid, just as in the case of a cumulative average, or Wright learning curve fit. This becomes more apparent when observing the parameters-balance, as there are obvious constraints which show-up in the interplay of parameter (a) and (b). When the first block data is absent from the data sets array to be used for the regression fit, substituting the data pairs, [100%; 0.5], creates an accurate surrogate that will maintain parameter balance.

For this example, the first data block is available, and is the same data used in CASE #2. Added to the first block are four follow-on blocks, from the source identified earlier.

Table (1) shows the proper IRATE data arrangement. The LCF is the slope incurred as a result of successive block accumulations, the learning curve computation re-computed with every block addition. The resulting cumulative average slopes are paired-off against the $[C / (C + R_n)]$ transform. The regression is calculated over the five data pairs. Figure (1) shows how the fitted five-block IRATE compares with the single block IRATE model of CASE #2.

Table 1. The Least Squares Fit of Actual Data Sets for Five Follow-on Blocks.

Block #	LCF	C/(C + R)	Actual Rate/Year
1st	96.742	0.5	Let R = C = 141
2nd	93.252	0.3294	287/Year
3rd	89.453	0.2427	440/Year
4th	86.986	0.1942	585/Year
5th	84.932	0.1475	815/Year

Fit Least Squares:
$$[C/(C + R_n)](33.595992) + 80.773397 = \hat{LCF}$$

ELEVEN MISSILE PROGRAMS

Missile program cost data is in sufficient supply, providing a variety of size, complexity, and programmatic backgrounds. Eleven programs, at various states of program maturity, were selected from the MISSILE DATA BOOK. IRATE Analysis was performed on each program, the results of which are shown in Table (2). Table (3) was formulated from Table (2), and includes those programs where all data sets were used in each IRATE assessment. There was one exception, the Harpoon program, where only the first five blocks were used. From Table (3) a generic missile production rate model emerges, formulated from the smear of averaging techniques displayed. The most significant revelation from the analysis is the C parameter. Like the B "expanse" parameter of the Stanford Model, C becomes a generic indicator of production type. Missile products have high pilot production rates, as the sample shows, causing a C parameter from 53 to 89 units. Figure (2) graphically depicts high and low averages as a spread band created from Table (3) data, for an idealized missile IRATE signature.

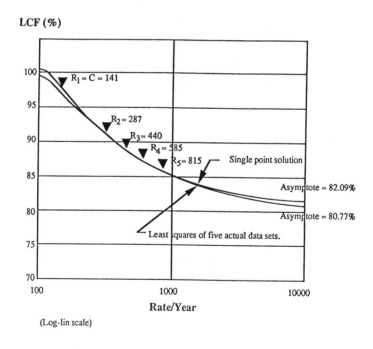

Figure 1. Actual Missile Program Learning Curve Progression.

Table 2. *IRATE Parameter Comparisons for Eleven Missile Programs.*

Missile Program		Highest Correlation	b	a	c	LCF at C
HARM	(All Data)	.7197	1.953	77.704	80	78.838
Harpoon	(First 5)	.6919	39.215	74.727	150	94.053
Hellfire	(High Rate)	.9998	10.975	81.548	3971	87.031
IHAWK	(4 late Blocks)	.9925	5.466	97.162	394	99.893
MAV65D	(No Pilots)	2 sets	17.024	83.184	3335	91.696
MAV65E	(No Pilots)	2 sets	20.720	85.984	393	96.344
MLRS	(High Rate)	.9999	38.135	70.476	36000	89.564
Patriot	(All Data)	.9930	37.167	78.568	176	96.742
PeaceMX	(All Data)	.9996	35.100	74.965	21	92.500
Pershing	(All Data)	.8417	14.185	88.687	70	95.761
Phnx. C	(All Data)	.9770	33.422	79.303	72	96.350

Table 3. *Developing a Generic Missile IRATE Model.*

Missile Program		Highest Correlation	b	a	c	LCF at C
HARM	(All Data)	.7197	1.953 *	77.704	80	78.838 *
Harpoon	(First 5)	.6919	39.215	74.727	150	94.053
Patriot	(All Data)	.9930	37.167	78.568	176 *	96.742
PeaceMX	(All Data)	.9996	35.100	74.965	21	92.500
Pershing	(All Data)	.8417	14.185	88.687 *	70	95.761
Phnx. C	(All Data)	.9770	33.422	79.303	72	96.350

Averaging the Parameters

	b	a	c	LCF at C
RMS:	30.203	79.130	108.3	92.583
Arithmetic Mean:	26.840	78.992	94.8	92.374
Geometric Mean:	19.022	78.861	77.9	92.148
Harmonic Mean:	8.654	78.736	59.6	91.904

Eliminating the outlier furthest from the mode (*):

	b	a	c	LCF at C
RMS:	33.074	77.076	88.8	95.094
Arithmetic Mean:	31.818	77.053	78.6	95.081
Geometric Mean:	29.989	77.031	66.2	95.068
Harmonic Mean:	27.577	77.008	52.7	95.055

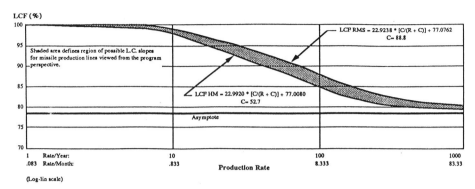

Figure 2. Generic Missile Rate Production Model.

LAUNCH VEHICLES

How closely related are launch vehicle IRATE signatures to those produced for tactical and strategic missiles? Considering the size of pilot production lots for missiles, it is not difficult to guess how the signature changes for the launch vehicle, keeping the same (a) and (b), and changing just C.

From a marketing source we find that one expendable launch vehicle currently reaches a procurement rate of five per year, while another obtains a rate more than double this. Comparing the two launch vehicle signatures with the missile signature finds the same general shape, but with improvements happening at steeper slopes for the launch vehicle lower production rates: 88% for the C = 5 assumption and 92% for the C = 12 assumption. This same production rate regime for the missile production line shows slopes in the mid-90% range. Thus, one can accept this conclusion if in agreement with the idea that larger assemblies, having the greater CIC potential because of the additional manual labor, will reach their learning potential at lower production rates, than will smaller missile types. Figure (3) shows how launch vehicle IRATE profiles compare with missile IRATE profiles.

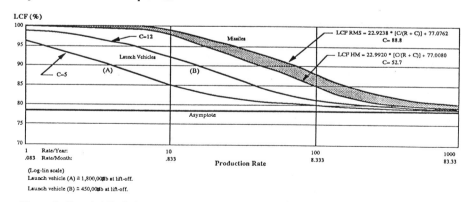

Figure 3. Generic Missile Rate Production Model Compared to Two Launch Vehicles LCF Profiles.

THE COEFFICIENT OF EFFICIENCY

Building the "grass roots" IRATE function is a challenge. There is a way by which the asymptote, or parameter (a), can be made into a detailed efficiency status device. By defining the entire production arena as a large two-dimensional layout of lines and stations, a chaining of efficiency ratings for each discrete element of the layout can result in a optimized CIC potential, which is in effect the target improvement objective, or the asymptote (a).

A detailed examination of this process is beyond the scope of this text. A brief synopsis of its features is presented to enable the analyst a full picture of IRATE capability, and at least an understanding of its potential applications.

Lines and Stations as a Two-dimensional Concept:

Using the Anderlohr [1] methodology for evaluating the elements of learning, it is possible to define a two-dimensional model for containing measures of element efficiency. Efficiency assignments for line m, station n, can be generated for the following:

Measures of Efficiency

C_{nm}	=	Continuity	(0<1)
P	=	Personnel	(0<1)
S	=	Supervision	(0<1)
M	=	methods	(0<1)
T	=	Tooling	(0<1)
A	=	Automation	(1 - % persons relieved by automation function)

The matrix of efficiency measures is combined and summed as follows:

$$p = \sum_{1}^{n} \sum_{1}^{m} C_{nm} \, P \, S \, M \, T \, A$$

Calculating the Asymptote (a) from the Efficiency Product:

Depending on the size of the Lines/Stations matrix, p, the coefficient of efficiency, may represent a number between 0 and infinity. The conversion of this number into a CIC asymptote is made possible by the equation illustrated in Figure (4). The function defines a CIC slope range between zero (no learning, 100%) and -1 (extreme learning, 50%). While the greatest band of variability occurs between $p = 0$ and $p = 10$, the largest band of slope constancy occurs between $p = 100$ and $p = 10$ exp 7, where the efficiency asymptote becomes 77.5%.

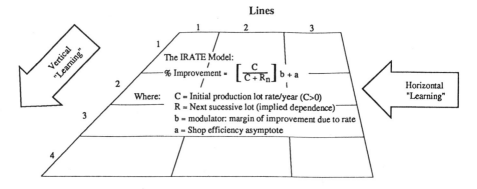

Suppliers "feed" the
lines/station matrix

People & ideas "pollenate"
the lines/station matrix

Lines

Vertical "Learning"

Horizontal "Learning"

The IRATE Model:

% Improvement = $\left[\dfrac{C}{C + R_n} \right]^{b} + a$

Where: C = Initial production lot rate/year (C>0)
R = Next sucessive lot (implied dependence)
b = modulator: margin of improvement due to rate
a = Shop efficiency asymptote

- **Efficiency Convergence Function:**

$$\% \text{ Shop Efficiency (A)} = \exp\left\{ \ln 2 * \left[-(1 - \frac{1}{1+p})^{1+p} \right] \right\} * 100$$

Find a continuous, converging series/function which will span the working range of improvement curve exponents (s), in the statement $y_c = T_1 * x^s$

Ones possible solution:

$\text{LIM} -[(1 - 1/1+p)^{1+P}] = -1$

$P \rightarrow \infty$

s = 0

100%

Y_c

s = -1

Learning
slope
band

X $\longrightarrow$

50%

- More than 1/3 of this efficiency slope (a) range (between 0 and -1) is scaled between p = 0 and p = 10
- The largest band of slope constancy is between p = 100 and p = 10^7, approximately 77.5%

Lines = m

Stations = n

$p = \sum\limits_{i=1}^{n} \sum\limits_{i=1}^{n} C_{nm} PSMTA$

Measure of Efficiency

C_{nm} = Continuity (0<1)
P = Personnel (0<1)
S = Supervision (0<1)
M = Methods (0<1)
T = Tooling (0<1)
A = Automation Fraction (1-% personnel relieved by automation function)

Figure 4. Creating the Efficiency Asymptote.

CONCLUSION

Most common misconceptions about production rate impact, are due to improper handling of the recurring and non-recurring parts of the problem, causing wrong conclusions about the relationship of cost increases or decreases incurred due to rate changes. The IRATE model handles the recurring problem, specifically the slope-altering characteristic. There is a significant relationship between production rate and the learning curve slope.

The IRATE model can be used in different ways. For the advanced programs estimating task IRATE can measure a single slope describing a nominal production rate following a Full Scale Development, or Pilot Production program rate. As a block-by-block procurement Tool, IRATE can measure the slope changes with every change of the production or procurement rate. The model will also measure program-end "toe-up" conditions.

EER has programmed the model in several spreadsheet apllications. We have found the tool to be very handy for adjusting a known CIC slope at a given production rate, to a new CIC slope at a new production rate. It is also a means for determining a CIC slope on a new system when there are no analoous programs for comparison.

Finally, the IRATE model can accommodate a most detailed evaluation of efficiency, where the asymptote of the function is independently derived from a critical accounting of lines and stations learning element performance measures. It is a model recommended for all DoD product lines.

References

[1]   Anderlohr, George "What Production Breaks Cost", Industrial Engineering, September 1969.

[2]   Chalmers, George and DeCarteret, Norman "Relationship for determining the Optimum Expansibility of the Elements of a Peacetime Aircraft Procurement Program", Stanford Research Institute No. 144, USAF Air Material Command, 31 Dec 1949.

[3]   Nicholas, T.G. "Missile Data Book", Data Search 1968.

A Bang-Bang Approximation to the Solution of a Learning Augmented Planning Model

James R. Dorroh
Department of Mathematics, Louisiana State University,
Baton Rouge, LA 70803

Thomas R. Gulledge, Jr.
George Mason University, 4400 University Drive, Fairfax, VA 22030

Norman Keith Womer
Department of Economics and Finance, University of Mississippi,
University, MS 38677

ABSTRACT

In this paper we study production programs where a relatively small number of complex units are produced to contractual order, and unit costs fall over time in some systematic way. We call this situation made-to-order production. Aircraft production is a good example. We derive the time path of planned resource use and demonstrate how to obtain approximate solutions for these optimal trajectories. The use of the approximation avoids a "messy" nonlinear optimization problem while providing resource estimates.

1.0 INTRODUCTION

In our previous research we have developed learning augmented planning models for production and cost analysis [8]. The emphasis was on made-to-order production, a situation where a small number of units are produced over the life of a contracted order. The characteristics of this production environment place unusual demands on government planners and cost estimators. Researchers have tried to model this situation using learning curves, but the learning curve is not appropriate for several reasons. The first of these reasons was noted many years ago by Asher [1]. The learning curve assumes constant production rate, and it incorporates no facility for treating production rate as a decision variable.

Using learning curves to model production costs when production rate is changing can badly distort the meaning of cost increases. In the extremes, cost overruns on defense programs, for example, may be attributed to poor management when they are due to exogenous changes in production rate. Because of these policy implications, we believe that it is important to formulate simple models of the production process which can incorporate the effects of production rate changes, and which are consistent with the prudent management of scarce resources. Some of the frustration over the lack of such models is evident in the following quote from a recent government study [4]. "Unfortunately, there is no mathematical formula to indicate when the rate of production becomes uneconomic. Service program managers and contractors have their own ideas, which do not always agree. Nevertheless, the services have previously reported minimum economic rates for a number of systems."

2.0 RELATION TO PREVIOUS WORK

Our research (summarized in [8]) has been concerned with understanding what Mody [14] calls the "black-box of learning-by-doing." As noted by Mody, "learning in an organization can take place at many different levels." Our research is devoted to understanding the production process improvement aspect of learning, in particular in the production of large made-to-order items such as aircraft.

We do not try to model production line details; instead our models examine the time trajectories of key planning variables. For example, total cost, aggregate production rate, or aggregate resource requirements are variables with which we often are concerned. Our intent is similar to the objectives of other researchers (e.g., Gaimon [5,6]) in that we want to understand the process dynamics, and perhaps be able to define general trends or strategies that may lead to better ways of doing things.

We offer no justification for the learning-by-doing phenomenon or its importance. This has been documented extensively by us and others [3,7,15,17,18]. Much of this literature assumes that cost falls in a reasonably regular manner as cumulative output increases. In effect, increases in cumulative output are assumed to be a proxy for firm learning. In this paper we are interested in modeling the effects of devoting some resources explicitly to knowledge creation. We separate and model those resources devoted to knowledge creation and those devoted to the production of physical output, an approach that was outlined in general terms by Rosen [16]. Our work combines Rosen's theoretical ideas with others identified by Womer [20]. The result is an economic control theory formulation, similar in scope to models described in detail by other researchers [10,11,13].

3.0 PROBLEM STATEMENT

In a previous paper [9], we modeled the made-to-order production situation using a dynamic optimization model. The model is best understood if compared with an earlier model [19]. Unlike the model in [19], the model in [9] permits the firm to allocate resources between the production of output and the production of increments to knowledge. Thus this model explores Killingsworth's [12] "investment in training" model of the life cycle.

The model in [19] leads to a time path of production rate which always increases. This contrasts with the observed timepath of production rate on many made-to-order production programs. The model in [9] permits production rate to increase early in the program and then to level off to a constant production rate, thus conforming more closely to the timepaths observed in many made-to-oorder production programs. The firm makes decisions that influence production rate, and ultimately the resources that are devoted to the production of output and product specific knowledge. Unfortunately, the solution to the original problem presents a difficult numerical estimation problem. In the present paper we present an approximate solution to the same problem while noting additional modeling relationships and results. The reason for reexamining the model is to find a simpler solution; one that is easy to estimate and test. Therefore, the primary motivation for this research was one of practicality, but other interesting model relationships were discovered as a byproduct of the modeling effort.

4.0 THE MODEL

The model and its solution are repeated here for continuity. The details of the solution are presented in [9]. The following definitions relate to the model:

$q(t)$ = program production rate at time t,

$\ell(t)$ = experience (learning) rate at time t,

$x_q(t)$ = the use rate of the variable composite resource devoted to the production of output,

$x_\ell(t)$ = the use rate of the variable composite resource devoted to the production of knowledge,

$x(t)$ = $x_q(t) + x_\ell(t)$

$Q(t)$ = $\int_0^t q(\tau)d\tau$ = cumulative output at time t,

$L(t)$ = $\int_0^t \ell(\tau)d\tau$ = cumulative stock of knowledge at time t.

γ = a factor returns parameter,

β = a factor returns parameter,

α = a learning parameter,

δ = a learning parameter

C = variable program cost measured in variable resource units

T = the time horizon for the production program,

V = volume of output to be produced by time T,

a_1 = a constant,

a_2 = a constant.

This situation is one where knowledge and output are produced by two production technologies. Notice that the cumulative stock of knowledge influences both output rate and the knowledge creation rate. The core production functions are

$$q(t) = a_1 x_q^{1/\gamma}(t)L^\alpha(t), \tag{1}$$

and

$$\ell(t) = a_2 x_\ell^{1/\beta}(t)L^\delta(t). \tag{2}$$

The time horizon is sufficiently short so that cost may be measured in units of the variable composite resources. Therefore the firm's problem is to

$$\text{minimize } C = \int_0^T [x_q(t) + x_\ell(t)]dt, \tag{3}$$

subject to: $\quad q(t) = a_1 x_q^{1/\gamma}(t) L^\alpha(t)$, $\qquad\qquad$ (4)

$\quad \ell(t) = a_2 x_\ell^{1/\beta}(t) L^\delta(t)$, $\qquad\qquad$ (5)

$\quad Q(0) = 0$, $\qquad\qquad$ (6)

$\quad Q(T) = V$, $\qquad\qquad$ (7)

$\quad L(0) = 0$, $\qquad\qquad$ (8)

$\quad L(T) = $ free. $\qquad\qquad$ (9)

The solution to this problem (see [9]) is of the following form:

$$\frac{t}{T} = \frac{\int_0^{Z^{-\eta}(T)Z^\eta(t)} y^{1/\eta-1}(1-y)^{(1-1/\beta)-1} dy}{\int_0^1 y^{1/\eta-1}(1-y)^{(1-1/\beta)-1} dy},$$ (10)

and

$$\frac{Q(t)}{V} = \frac{\int_0^{Z^{-\eta}(T)Z^\eta(t)} y^{(1/\eta-1)-1}(1-y)^{(1-1/\beta)-1} dy}{\int_0^1 y^{(1/\eta-1)-1}(1-y)^{(1-1/\beta)-1} dy},$$ (11)

where $Z(t) = L^{1-\delta}(t)/(1-\delta)$, and $\eta = \alpha\gamma/[(1-\delta)(\gamma-1)]$.

Equations (10) and (11) provide a convenient way to state the solution in terms of incomplete beta functions if the model's parameters are known. Although we can a priori provide aproximate values, these parameters should be estimated from actual weapon system program data. The solution to equations (10) and (11) in that case represents a very complex and time consuming parameter estimation problem. In the next section, additional model relationships are derived that lead to an approximate solution to this problem.

5.0 ADDITIONAL MODEL RELATIONSHIPS

The resource requirement functions are defined by solving the core functions [equations (1) and (2)] for the variable composite resources. The resulting expressions are

$$x_q(t) = a_1^{-\gamma} q^\gamma(t) L^{-\alpha\gamma}(t),$$ (12)

and

$$x_\ell(t) = a_2^{-\beta} \ell^\beta(t) L^{-\delta\beta}(t).$$ (13)

The intermediate function for the optimization is $I = x_q(t) + x_\ell(t)$, and the Euler equations are

$$\frac{d}{dt}[I_q] = I_Q,$$ (14)

and

$$\frac{d}{dt}[I_\ell] = I_L.$$ (15)

In equations (14) and (15), the subscript notation indicates partial differentiation. A special form of the Euler equation [2, p. 15] requires that

$$\frac{d}{dt}[I-q(t)I_q-\ell(t)I_\ell] = \frac{\partial I}{\partial t} = 0. \tag{16}$$

Now

$$I_q = \frac{\partial x_q}{\partial q} = \frac{\partial}{\partial q}[q^\gamma a_1^{-\gamma}L^{-\alpha\gamma}] = \gamma q^{\gamma-1}a_1^{-\gamma}L^{-\alpha\gamma}. \tag{17}$$

This implies that

$$q(t)I_q = a_1 x_q^{1/\gamma}(t)L^\alpha(t)\gamma q^{\gamma-1}(t)a_1^{-\gamma}L^{-\alpha\gamma}(t) = \gamma x_q(t). \tag{18}$$

By a similar sequence of steps it follows that

$$\ell(t)I_\ell = \beta x_\ell(t). \tag{19}$$

If equations (18) and (19) are substituted into equation (16), the following expression is obtained:

$$\frac{d}{dt}[(1-\gamma)x_q(t) + (1-\beta)x_\ell(t)] = 0. \tag{20}$$

Equation (20) integrates to a constant; that is, the following linear relationship exists between the resources:

$$(1-\gamma)x_q(t) + (1-\beta)x_\ell(t) = K \tag{21}$$

where K is a constant. This fundamental restriction on our model also holds on the approximate solution in the next section, even though the approximation is the result of an independent derivation.

6.0 THE BANG-BANG APPROXIMATION

In this section we approximate the resource extremals with a bang-bang solution to the problem. For the parameter ranges of interest, the optimal strategy in all cases is to devote resources to knowledge early in the program. After this initial period of knowledge creation, the solution requires that knowledge production cease, and resources at that point in time are devoted to output production. This type of behavior is often encountered on programs where a small number of complex units are produced to contractual order. Therefore, in this section, we will minimize total cost for a restricted class of models. In this restricted class, all resources will be devoted to learning early in the program; then at a certain point, resources are no longer allocated to learning, and all resources are allocated to output. If this model is to satisfy equation (21), then x_ℓ should have a constant value, A, on the initial time interval, and x_q should have a constant value, B, on the complementary interval. Furthermore, equation (21) implies that $(1-\gamma)A = (1-\beta)B$, but in this section this relationship is derived independently.

The following notation is used for the bang-bang formulation. Let

$$X_E(t) = \begin{cases} 1 & \text{when t is an element of E,} \\ 0 & \text{otherwise.} \end{cases} \tag{22}$$

Thus, the resource requirement functions are given by

and
$$x_q = {}^{BX}[T_1,T],$$ (23)

$$x_\ell = {}^{AX}[0,T_1].$$ (24)

Begin the solution by examining equation (2), the knowledge production function. From equations (2) and (24) we obtain

$$L^{-\delta}(t)\ell(t) = a_2 A^{1/\beta}.$$ (25)

If we integrate both sides of equation (25) from zero to T_1 and solve for $L(T_1)$, we obtain

$$L(T_1) = C_1 A^{1/\beta(1-\delta)} T_1^{1/(1-\delta)}$$ (26)

where $C_1 = (1-\delta)^{1/(1-\delta)} a_2^{1/(1-\delta)}$. In fact, since $x_\ell(t)$ vanishes for t in $[T_1,T]$, then, by equation (2), $\ell(t)$ vanishes as well. Therefore

$$L(t) = C_1 A^{1/\beta(1-\delta)} T_1^{1/(1-\delta)}$$ (27)

for all t in $[T_1,T]$. Now examine equation (1), the output production function. If equation (26) is substituted in equation (1), the following expression for output rate is obtained:

$$q(t) = C_2 B^{1/\gamma} A^{\alpha/\beta(1-\delta)} T_1^{\alpha/(1-\delta)}$$ (28)

for $T_1 \le t \le T$, where $C_2 = a_1 C_1^{\alpha}$, and B is the value for $x_q(t)$ given in equation (23). To obtain an expression for cumulative output, integrate both sides of equation (28) with respect to t and obtain

$$Q(T) = C_2 B^{1/\gamma} A^{\alpha/\beta(1-\delta)} T_1^{\alpha/(1-\delta)} (T-T_1).$$ (29)

The objective function for minimization is given by $C = AT_1 + B(T-T_1)$, and we know $Q(T) = V$.

This leads to the following optimization problem:

minimize $C = AT_1 + B(T-T_1),$ (30)

subject to: $Q(T) = V.$

The problem may be solved by optimizing the Lagrangian function,

$$H = AT_1 + B(T-T_1) + \lambda[V-Q(T)].$$ (31)

The necessary conditions are

$$\frac{\partial H}{\partial A} = T_1 - \lambda \frac{\partial Q(T)}{\partial A} = 0,$$ (32)

$$\frac{\partial H}{\partial B} = T - T_1 - \lambda \frac{\partial Q(T)}{\partial B},$$ (33)

$$\frac{\partial H}{\partial T_1} = A - B - \lambda \frac{\partial Q(T)}{\partial T_1} = 0,$$ (34)

$$\frac{\partial H}{\partial \lambda} = V - Q(T) = 0.$$ (35)

Equation (29) may be used to obtain the relevant derivatives; that is,

$$\frac{\partial Q(T)}{\partial A} = \alpha B^{-1}(1-\delta)^{-1}C_2 B^{1/\gamma}A^{[\alpha/\beta(1-\delta)]-1}T_1^{\alpha/(1-\delta)}(T-T_1).$$

$$= \alpha Q(T)B^{-1}(1-\delta)^{-1}A^{-1},$$ (36)

and

$$\frac{\partial Q(T)}{\partial B} = \gamma^{-1}C_2 B^{(1/\gamma)-1}A^{\alpha/\beta(1-\delta)}T_1^{\alpha/(1-\delta)}(T-T_1) = Q\gamma^{-1}B^{-1},$$ (37)

and

$$\frac{\partial Q(T)}{\partial T_1} = \alpha(1-\delta)^{-1}C_2 B^{1/\gamma}A^{\alpha/\beta(1-\delta)}T_1^{[\alpha/(1-\delta)]-1}(T-T_1)$$

$$- C_2 B^{1/\gamma}A^{\alpha/\beta(1-\delta)}T_1^{\alpha/(1-\delta)}$$

$$= Q[\alpha(T-T_1)-(1-\delta)T_1](1-\delta)^{-1}T_1^{-1}(T-T_1)^{-1}$$ (38)

We obtain the following equations:

$$\beta A = \lambda \alpha V T_1^{-1}(1-\delta)^{-1},$$ (39)

$$\gamma B = \lambda V(T-T_1)^{-1},$$ (40)

$$A-B = \lambda V[\alpha(T-T_1)-(1-\delta)T_1](1-\delta)^{-1}T_1^{-1}(T-T_1)^{-1}.$$ (41)

Equation (39) follows from equations (32), (36), and (35). Equation (40) follows from equations (33), (37), and (35). Equation (41) follows from equations (34), (38), and (35). If equation (40) is subtracted from equation (39), the following expression is obtained:

$$\beta A - \gamma B = \lambda V[\alpha(T-T_1)-T_1(1-\delta)](1-\delta)^{-1}T_1^{-1}(T-T_1)^{-1}.$$ (42)

Since the right hand sides of equations (41) and (42) are the same, we obtain

$$A(\beta-1) = B(\gamma-1)$$ (43)

which is the same linear relationship between the resources that was derived by the calculus of variations solution to the problem.

To solve for T_1, the time at which knowledge production ceases and output production begins, follow this sequence of steps. Solve equation (39) for T_1 and equation (40) for $T-T_1$. Take the ratio of the resulting expressions to obtain

$$T_1/(T-T_1) = (B/A)[\alpha\gamma\beta^{-1}(1-\delta)^{-1}]. \tag{44}$$

But equation (43) implies $B/A = (\beta-1)/(\gamma-1)$, therefore

$$T_1/(T-T_1) = \alpha\gamma(\beta-1)\beta^{-1}(1-\delta)^{-1}(\gamma-1)^{-1} = r. \tag{45}$$

This implies

$$T_1 = [r/(1+r)]T \tag{46}$$

Where r is defined in equation (45).

To find the solutions for A and B, use the last condition, $Q(T)=V$. Equation (29) implies

$$C_2 B^{1/\gamma} A^{\alpha/\beta(1-\delta)} T_1^{\alpha/(1-\delta)}(T-T_1) = V, \tag{47}$$

and equation (43) implies

$$A = B(\gamma-1)/(\beta-1). \tag{48}$$

Use equation (48) to eliminate A from equation (47), and solve for B. The expression is

$$B = C_3 T_1^{-\alpha/[\Omega(1-\delta)]}(T-T_1)^{-1/\Omega} \tag{49}$$

where $C_3 = V^{1/\Omega} C_2^{-1/\Omega}[(\gamma-1)/(\beta-1)]^{-\alpha/\beta\Omega(1-\delta)}$, and $\Omega = [\beta(1-\delta)+\alpha\gamma]/\gamma\beta(1-\delta)$.

Equation (48) may now be used to obtain an expression for A.

The solution is now complete. The results for the variables of interest are summarized below:

$$T_1 = [r/(1+r)]T, \tag{50}$$

$$A = C_3(\gamma-1)(\beta-1)^{-1}T_1^{-\alpha/[\Omega(1-\delta)]}(T-T_1)^{-1/\Omega}, \tag{51}$$

$$B = C_3 T_1^{-\alpha\Omega/(1-\delta)}(T-T_1)^{-1/\Omega}, \tag{52}$$

$$C = AT_1 + B(T-T_1). \tag{53}$$

These relationships are used to approximate the solution to the problem defined in equations (3) through (9). The implementation of the approximation is analyzed in the next section.

7.0 IMPLEMENTATION

The implementation of this model is simple. Three estimates are needed; estimates for A, B, and T_1. The parameter A is associated with all the cost data points that occur prior to time T_1, and B is associated with those cost data points that occur after T_1. It is easy to show that the estimates for A and B that minimize the sum of squared errors are just the averages of the data points that occur before and after T_1, respectively. This result is rather intuitive, but an estimate for T_1 is still required. Since the data is time sequenced, and since made-to-order programs have relatively short time horizons, T_1 may be estimated by enumeration. That is, the sum of squared error function is computed for all possible values of T_1, and the optimal value (the one that minimizes the squared error) is selected.

8.0 CONCLUSION

This paper presents an approximate solution to a learning augmented planning problem, a problem that arises in made-to-order production. The solution is simple, and the model may be used to easily provide approximate solutions to optimal resource time paths, and to indentify the optimal time to divert resources from the production of knowledge to the production of output.

9.0 REFERENCES

[1] Asher, Harold. Cost-Quantity Relationships in the Airframe Industry, R- 291. Santa Monica: The RAND Corporation, 1956.

[2] Bliss, Gilbert A. Lectures on the Calculus of Variations. Chicago: The University of Chicago Press, 1946.

[3] Camm, Jeffrey D.; Thomas R. Gulledge, Jr.; and Norman Keith Womer. "Production Rate and Optimal Contractor Behavior," The Journal of Cost Analysis, Vol. 5 (1987), 27-37.

[4] Congressional Budget Office. Effects of Weapons Procurement Stretch-Outs on Costs and Schedules. Washington D.C.: Congress of the United States, November 1987.

[5] Gaimon, Cheryl "Simultaneous and Dynamic Price, Production, Inventory and Capacity Decisions," European Journal of Operational Research, Vol. 35 (1988), 426-441.

[6] Gaimon, Cheryl "The Optimal Acquisition of Automation to Enhance the Productivity of Labor," Management Science, Vol. 31 (1985), 1175-1190.

[7] Gulledge, Thomas R., Jr. and Bekrokh Khoshnevis. "Production Rate, Learning, and Program Costs: Survey and Bibliography," Engineering Costs and Production Economics, Vol. 11 (1987), 223-236.

[8] Gulledge, Thomas R. and Norman Keith Womer. The Economics of Made-to-Order Production. Berlin: Springer-Verlag, 1986.

[9] Gulledge, Thomas R., Jr.; Norman K. Womer; and James R. Dorroh. "Learning and Costs in Airframe Production: A Multiple Output Production Function Approach," Naval Research Logistics Quarterly, Vol. 31 (1984), 67-85.

[10] Intriligator, Michael D. Mathematical Optimization and Economic Theory. Prentice-Hall: Englewood Cliffs, 1971.

[11] Kamien, Morton I. and Nancy L. Schwartz. <u>Dynamic Optimization: The Calculus of Variations and Optimal Control in Economics and Management</u>. New York: North-Holland, 1981.

[12] Killingsworth, Mark R. "Learning by Doing and Investment in Training: A Synthesis of Two Rival Models of the Life Cycle," <u>Review of Economic Studies</u>, Vol. 49 (1982), 263-271.

[13] Miller, Ronald E. <u>Dynamic Optimization and Economic Applications</u>. New York: McGraw-Hill, 1979.

[14] Mody, Ashoka. "Firm Strategies for Costly Engineering Learning," <u>Management Science</u>, Vol. 35 (1989), 496-512.

[15] Muth, John F. "Search Theory and the Manufacturing Progress Function," <u>Management Science</u>, Vol. 32 (1986), 948-962.

[16] Rosen, Sherwin. "Learning by Experiene as Joint Production," <u>Quarterly Journal of Economics</u>, Vol. 86 (1972), 366-382.

[17] Towill, Denis R. "Management Systems Applications of Learning Curves and Progress Functions," <u>Engineering Costs and Production Economics</u>, Vol. 9 (1985), 369-383.

[18] Venezia, Itzhak. "On the Statistical Origins of the Learning Curve," <u>European Journal of Operational Research</u>, Vol. 19 (1985), 191-200.

[19] Womer, Norman Keith. "Learning Curves, Production Rate, and Program Costs," <u>Management Science</u>, Vol. 25 (1979), 312-319.

[20] Womer, Norman Keith. "Some Propositions on Cost Functions," <u>Southern Economic Journal</u>, Vol. 47 (1981), 1111-1119.

II. Software Economics

The Software Development Effort Estimation Exponent: An Effort Estimation Model Accounting for Learning and Integration

Everett Ayers

3203 Harness Creek Road, Annapolis, MD 21403

INTRODUCTION

The study of estimation of the effort required for software development has progressed to the point where most currently used models employ
an equation of the form $y = ax^e$, which relates the effort to the lines of
code, or size of a program, exponentially through a constant, e. That
constant, e, the effort estimation exponent, has been referred to as an
entropy constant by Randall Jensen[1] and is the subject of many analyses,
both theoretical and empirical, to estimate its value. In fact, many
models indicate that its value is greater than one, meaning that the
effort estimation curve bends upward, or that larger programs take
proportionally more effort to develop. Yet, some estimates of the
exponent's value are less than one, and some SEL experience indicates a
value of 0.92 and a curve that bends downward. This implies that larger
programs take proportionally less effort to develop. Furthermore, if an
exponent greater than one is employed, then the effort estimated to
develop a large program will be greater than the sum of the effort
required to develop each of its submodules individually. In this paper,
learning curves and integration effort estimators are applied to explain
why the effort estimation curve might bend upward or downward for specific applications and why integration effort should be estimated as part
of every software development, including the effects of reusable code,
if it applies. Also, this paper presents the development of a model
to incorporate learning curves and integration into the exponent and
therefore to estimate the quantitative effects of integration and
learning, and to explain why and when the effort estimation curve
bends upward or downward.

---

[1]R. W., Jensen, "Sensitivity Analysis of the Jensen Software Model,"
and "A Macro-Level Software Development Cost Estimation Methodology,"
Fourteenth-Asilomar Conference on Circuits, Systems and Computers,
IEEE, New York, 1981.

Two ideas guided this study, therefore. First, the recognition that, if the exponent is greater than one, then $I^e > I_1^e + I_2^e ... + I_n^e$ where $I = \sum_{i=1}^{n} I_i$, and this provides an estimator for integration effort and an explanation for an upward curve. Secondly, the parallels with learning curves provided an explanation for a downward curve and quantitative estimates of learning effects.

BACKGROUND

Software development effort estimation equations in general use currently take an exponential form such as:

$$MM = PI^e \qquad (1)$$

where MM = Effort, in manmonths
       P = Productivity, in man-months per thousand lines of code
       I = Source Lines of Code, in thousands of lines
       e = Exponent, a constant depending upon application and
            environment.

The Productivity is often modified by multiplying it by several factors affecting productivity such as language, experience, environment, and resources. This study will concentrate on an examination of the nature of the exponent because it determines the shape of the curve and because it can be employed to estimate learning and integration effects.

The form of the software effort estimation equation has evolved from its early linear form to the exponential form above. For historical perspective, several models will be briefly summarized to trace the development of effort estimation equations. The 1965 SDC Model[2] was a linear statistical model based upon project attributes and not upon lines of code. In 1969, Aron[3] of IBM introduced a linear equation for software development man-months of the form:

$$MM = SHM \ I/P \qquad (2)$$

---

[2]E. A. Nelson, "Management Handbook for the Estimation of Computer Programming Costs" AD-A648750, Systems Development Corp., Oct 31, 1966.

[3]J. D. Aron, "Estimating Resources for Large Programming Systems," NATO Science Committee, Rome, Italy, October 1969.

where P, the productivity factor, depended on the project duration and the project difficulty. In 1974, Wolverton[4] developed an estimating equation of similar form, but extended the concept to include the productivity factor being a function of the application categorical type. For ESD in 1977, Doty Associates[5] developed a number of statistical, regression-based models of the exponential form:

$$MM = aI^b \qquad (3)$$

where a ranged from 2.9 to 12.0 and b ranged from 0.719 to 1.263, depending upon the application type. This was important in the introduction of the exponential form. Also in 1977, Walston and Felix[6] of IBM developed a software cost estimating relationship (CER) as follows:

$$MM = 5.2L^{0.91} \qquad (4)$$

where L is delivered source code in thousands of lines and the exponent is less than one. These were all empirically based models.

In the same time period, Norden[7] of IBM proposed the use of the Rayleigh equation for staffing levels of research and development projects consisting of overlapping work cycles. Putnam[8] applied the Rayleigh-Norden equation to USACSC software data and developed mathematical relationships for a macro-estimation model. The model implies an effort-size relationship of the exponential form:

$$E = aS^\alpha \qquad (5)$$

where α is 1.263. The Jensen[1] macrolevel software development resource estimation model evolved from the above work by modifying the productivity slope of Putnam's data to -0.50 because of a refined data set. The modification leads to the mathematical conclusion than the exponent in

[4]R. W. Wolverton, "The Cost of Developing Large-Scale Software, "IEEE Transactions on Computers, pp. 615-636, June 1974.

[5]Doty Associates, Inc., "Software Cost Estimation Study, Guidelines for Improved Software Cost Estimating," Volume II, RADC-TR-77-220, August 1977.

[6]C. E. Walston and C.; P. Felix, "A Method of Programming Measurement and Estimation," IBM System Journal, 16, 1, 1977.

[7]Norden, P. V., "Project Life Cycle Modeling: Background and Application of Life Cycle Curves," Software Life Cycle Management Workshop, USACSC, Airlie, VA, August 1977.

[8]Putnam, L. H., "Example of an Early Sizing, Cost, and Schedule Estimate for an Application Software System," COMPSAC 1978, November 13/16 1978.

the effort size relationship is 6/5, or 1.20. Jensen's model includes
the effective software size, S_e, which counts not only the newly developed source code but also accounts for the interaction between source
code modifications and the existing software. The extra effort required
to incorporate new or modified software is related to design, test, and
code interface; and effective size is greater than the actual size of the
modification. The solutions obtained from the Jensen model represent 50
percent probability values of schedule and cost. The Putnam and Jensen
formulations are mathematical and derivative in nature, based upon equations and assumptions, with empirical data applied to evaluate constants
and effect calibration.

In 1981, Dr. Barry Boehm published the COnstructive COst MOdel[9]
(COCOMO), a detailed, empirical formulation which presented effort-
size relationships, for the Basic and Intermediate models, as

$$MM = 2.4 \ KDSI^{1.05} \quad \text{(Organic Mode)} \tag{6}$$
$$MM = 3.0 \ KDSI^{1.12} \quad \text{(Semidetached Mode)}$$
$$MM = 3.6 \ KDSI^{1.20} \quad \text{(Embedded Mode)}$$

Many cost driver attributes are also presented to fine-tune the basic
estimates. The Basic COCOMO estimating equations were obtained from
analysis of a carefully screened sample of 63 software project data
points. Thus, the exponents, which are greater than one, have been
derived empirically.

The NASA/GSFC Software Engineering Laboratory (SEL) found through its
experience that the exponent of the model $E = aL^b$ could be less than one,
0.92, in fact, as shown in Figure 1, despite the popular assumption that
it was greater than one.

Therefore, as Table 1[1] illustrates, the value of the software effort
estimation exponent has been modeled mathematically and empirically and
estimated to be greater than one in some models and less than one in
other models. How can the same curve bend both up and down? This paper
presents the viewpoint that it can bend either upward or downward,
depending upon the application and the resources applied, and also
presents a model for estimating what the exponent will be as a function
of the application and the resources.

---

[9]Boehm, B. W., _Software Engineering Economics_, Prentice-Hall, Inc.,
Englewood Cliffs, NJ, 1981.

TABLE 1

SOFTWARE ESTIMATION MODELS AND EXPONENTS

MODEL	EXPONENT
Aron	1.00
Wolverton	1.00
Doty Associates	0.781 to 1.263
Walston and Felix	0.91
Putnam SLIM	1.263
Jensen SEER	1.20
COCOMO Organic	1.05
COCOMO Semidetached	1.12
COCOMO Embedded	1.20
Halstead	1.50
SEL (Figure 1)	0.92

LEARNING CURVES

Early software development effort estimation models presumed that code production was linearly related to the number of lines of code by productivity constants and other multipliers. That is, the effort estimation exponent is equal to one. This is a very reasonable, first assumption, and, if software lines of code were written by a perfect, uniform, memoryless code-writing machine turning out n number of lines per day constantly, it would be exactly correct. But software engineers and programmers writing code do have memory, do learn from experience, can reuse code or portions thereof, do have specialties and do have company resources and software tools to help them; basically, they learn from experience, and it is reasonable to suppose that learning experience can apply on large programs or from program to program. Learning experience based upon individual experience and capability, as well as corporate memory and experience for a similar applications would explain why a software effort curve could bend downward in certain cases. Learning curves are, therefore, useful for predicting when and how much the effort curve bends downward. They provide interesting parallels with manufacturing and with reliability. Many experts, including the SEL, believe that people are the most important resource/technology in software engineering. The application of learning curves to the effort estimation exponent allows for an accountability of the advantages of using more

FIGURE 1. ARE LARGE PROGRAMS HARDER
TO BUILD THAN SMALL ONES?

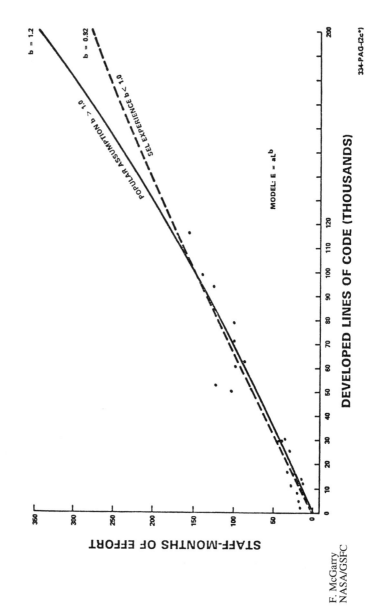

F. McGarry
NASA/GSFC

experienced, better people to produce code more effectively and proportionately more efficiently on a large project.

Learning curves were first employed in the aircraft industry to forecast the effect of learning on production rate.[10] Data from World War II supported a learning curve equation of the form:

$$Y = Ax^b \tag{7}$$

where Y = Average cost per unit
A = First Unit Cost
x = Number of units
b = Learning curve slope parameter.

That is, production efficiency increases logarithmically with respect to production quantity. The value of b is negative, defined by

$$b = \ln f / \ln 2 \tag{8}$$

where f is the learning factor, a value between 0 and 100%. The rate of improvement, or slope, of the learning curve relates the percentage of improvement for each doubling of production quantity, because of the ln 2 factor, and is a constant. For example, in a 90% learning curve, doubling the production quantity reduces average unit labor to 90%. Since the original applications to aircraft assembly, the learning curve has been found to fit a variety of operations. Different kind of operations tend toward different slope values, and learning factors can be estimated from the content of an operation. The learning curve is useful for many functions showing exponential decay. The reliability growth model used by Duane[11] and others is identical to the learning curve relationship. The Duane model is equivalent to a Weibull stochastic process with a specific intensity function. Parallels between learning curves for manufacturing and reliability growth modeling are illustrated in Table 2.

[10]Barton, Jr., H. R., "Predicting Guaranty Support Using Learning Curves," Proceedings of the Annual Reliability and Maintainability Symposium, IEEE, 1985.

[11]Duane, J. T., "Learning Curve Approach to Reliability Monitoring," IEEE Transactions on Aerospace, Vol. 2, No. 2, April 1964.

TABLE 2

PARALLELS IN LEARNING CURVE AND RELIABILITY GROWTH PARAMETERS

Manufacturing Learning Curve Parameter	Reliability Growth Parameter
Average Cost per Unit	Cumulative Failure Rate
First Unit Cost	Initial Failure Rate
Number of Units Produced	Cumulative Operating Hours
Slope Parameter	Slope Parameter

For total effort estimation, the learning curve equation becomes, by multiplying through by number of units:

$$Cost \ = \ Ax^{1+b} \tag{9}$$

where b = ln f/ln 2 and is negative since f is between 0 and 1. The analogy to software effort estimation is:

$$MM \ = \ PI^{1+b} \tag{10}$$

where MM = man-months,
      P = productivity constant in man-months/KSLOC,
      I = KSLOC,
and b = learning curve parameter, as above.

This equation explains how the software effort estimation curve can bend downward. For example, an exponent of 0.92, as suggested in Figure 1, relates to a 95% learning curve parameter. The slope parameter is -0.152 for 90%, -0.234 for 85%, and -0.322 for 80%.

Practical considerations and the relatively little applicable software experience possible relative to the extent of manufacturing experience indicate that the learning curve for software is typically 90% and perhaps 80% at an extreme, but seldom less than 80%. If b is related to programmer learning and experience and corporate memory relative to similar applications, then a simple scale of years of experience may be practical to estimate the learning curve parameter, whereas in actuality and in concept, the parameter depends on many, more complex factors including the application and perhaps even the size. At first examination for simplicity and practicality, a scale such as in Table 3 could be a reasonable approximation dealing with the average number of programmer years of experience. The learning curve factor, in principle, would differ by company, by application,

TABLE 3

SOFTWARE LEARNING CURVE PARAMETERS

Average Number of Years Experience	Learning Curve Factor	Value of b
0-1⁻	95%	-0.074
1⁺-3⁻	90%	-0.152
3⁻-5	85%	-0.234
over 5	80%	-0.322

and by language. b could be a continuous variable, and should undergo statistical calibration with extensive software project data. The calibration, however, should also include the effects of integration on the exponent, together with learning, as discussed below.

SOFTWARE INTEGRATION

The learning curve effect explains an effort exponent less than one and a downward curve, but does not explain exponents greater than one and upward curves. There must be at least one more effect. A clue to such an effect came when estimating effort for a large program consisting of n modules. Using a Basic COCOMO, organic mode exponent, 1.05, or any exponent greater than one in fact, it became apparent than writing a large program required more effort than writing n separate modules individually, even if the same number of SLOC were involved because

$$I^\alpha > I_1^\alpha + I_2^\alpha + \ldots + I_n^\alpha \qquad \text{for } \alpha > 1 \text{ and } I = \sum_{i=1}^{n} I_i \qquad (11)$$

The whole is greater than the sum of the parts, in this instance, and it exceeds the sum of the parts by more as α increases above 1.0. That is, if

$$MM_{TOTAL} = PI^{\alpha} \tag{12}$$

and

$$MM_i = PI_i^{\alpha},$$

then

$$MM_{TOTAL} = PI^{\alpha} > \sum_{i=1}^{n} MM_i = \sum_{i=1}^{n} PI_i^{\alpha} \qquad \text{for } \alpha > 1$$

A major reason why it takes more effort to write a large software program (of 100 K lines, for example, even in modules) than to write 100 programs of 1 K lines each is the interaction, interplay, and interfaces between and among the modules. Variables must be interfaced, common data must be shared, timing and order must be preserved, program flow must be cross-checked, etc. All of these difficult tasks comprise what this paper will refer to as software integration, or integration. Obviously, it takes much effort to coordinate the development, to perform these tasks, and to turn n individual modules into a comprehensive, functioning, whole software program. It is this extra effort that shows up if we subtract ΣMM_i from MM_{TOTAL} and get:

$$\Delta MM = P (I^{\alpha} - \Sigma I_i^{\alpha}) \qquad \text{for } \alpha > 1 \tag{13}$$

This paper proposes that this extra effort be referred to as software integration effort, that it is this effect which causes the effort estimation curve to bend upward, and that integration effort can be accounted for and estimated by the inclusion of an integration term in the effort estimation exponent. As the program size increases, so does the integration effort because of the greater number of interfaces, program paths, and interactions. Software integration effort relates to the extra effort to incorporate new or modified software through design, test, and code interface as described by Jensen.[1]

Because of the similarities to the learning curve term and because of its practicality, this paper presents a candidate estimator for the integration portion of the effort estimation exponent in logarithmic form. The range of historical estimates for the effort estimation exponent is from 0.71 to 1,263, or 1.50 (3/2) if Halstead's equation is included. With learning curve exponents, b, ranging from 0 to -.322, typically, it makes sense that the integration exponent should vary from

0 to approximately .6. If the relative difficulty of integration or the extent of integration can be quantified between 0 and 100% (denoted f_i), then the integration effort exponent, i, can be estimated by the function:

$$i = \frac{\ln(2 + f_i)}{\ln 2} - 1 \tag{14}$$

i ranges from 0 to 0.58. For a nominal 50% value of f_i, the relative integration difficulty, then the integration exponent, i, is 0.28. For a 30% integration difficulty, the integration exponent is 0.20, and the overall effort estimation exponent, combining learning and integration as 1 + b + i, is 1.20 + b, which is the COCOMO Embedded Mode exponent's and the Jensen exponent's value without learning.

Recognize that the integration difficulty percentage is, as yet, a very subjective quantity, which likely depends upon the application, the size, the complexity, the number of modules, the amount of reused code, and many other factors. Calibration and verification with actual data are the highly desirable next steps, which can lead to tables of sample values and characteristics. For now, it is possible to use subjective judgement around nominal 30% and 50% values to determine whether the integration tasks are more or less difficult than average. To be sure, a more quantitative measure is both desired and possible. One candidate based upon the Function Point Methodology[12] is derived by taking the ratio of Function Points directly related to integration areas; specifically, inquiries, files, and interfaces; to the total number of Function Points. That is, for average Function Point weighting factors:

$$f_i = \frac{4 \text{ (Inquiries)} + 10 \text{ (Files)} + 7 \text{ (Interfaces)}}{4 \text{ (Inputs)} + 5 \text{ (Outputs)} + 4 \text{ (Inquiries)} + 10 \text{ (Files)} + 7 \text{ (Interfaces)}} \tag{15}$$

is a quantitative estimate of the relative integration difficulty or effort.

The exponent, e = 1 + b + i, yields an estimate of integration effort for evaluation:

---

[12]Albrecht, A. J. and Gaffney, Jr., J. E., "Software Function, Lines of Code, and Development Effort Prediction: A Software Science Validation, IEEE Transactions on Software Engineering, November 1983.

$$MM_{Int} = PI_{TOTAL}^e - P \sum_{i=1}^{n} I_i^e \qquad \text{for } e > 1, \tag{16}$$

where the program is developed as n modules of I_i KSLOC to be integrated together. These values can be employed to estimate the amount of effort to be applied to (or to be required for) integration, a quantity useful in planning, budgeting, and staffing. Furthermore, with reusable code becoming more and more important and available, this estimating methodology provides a way to account for the effort required to integrate reusable code into a program when the code itself is not to be initially developed nor rewritten for the specific program, but only to be incorporated into a larger program. Then, the equation:

$$MM_{Int} = P (I_{reuse} + I_{new})^e - P (I_{reuse}^e + I_{new}^e) \qquad \text{for } e > 1 \tag{17}$$

can be employed to estimate the integration effort for reusable code. This knowledge could be critical to the decisions whether to use reusable code or rewrite, and b and i can be estimated by themselves or can be inferred based on a given value of the effort estimation exponent, $1 + b + i$.

Integration difficulties of nominally 50% and integration exponential factors of 0.32, combined with nominal 90% learning curves and learning exponential factors of -0.15, give an effort estimation exponent of 1.17. For illustration, a program of 10 K lines of code written as five modules of 2 KSLOC each would require that 7.8 out of 32.5 man-months, or 24% of the effort be devoted to "integration", at the optimum. It has long been the author's viewpoint that software estimates from the models can be construed as either optimum (best case) or most likely values. There is nothing to prevent an equally good program from being 120 K lines, for example, rather than 100 K lines, assuming enough memory space, but it is the functionality that can be achieved in 100 K lines relative to the functionality that can be achieved in programs of other size, at the optimum. Also, if the functionally can be achieved in 500 man-months and in 24 calendar months, that is certainly not to say it won't take longer, as software engineers know, but an optimum, or a nominal, standard for comparison is established. For example, the Jensen estimates of schedule and cost are defined as 50% probability values.

THE DEVELOPMENT SCHEDULE EQUATION

The software development models of Boehm, Putnam, Jensen, and Walston-Felix include expressions for the development time, T, as a function of the effort, E, of the form:

$$T = a E^e \tag{18}$$

where a varies from 2.29 to 4.48 for the models, and e varies from 0.32 to 0.35. In fact, mathematical derivations by Putnam and by Jensen indicate that the exponent is actually 1/3. Therefore,

$$T \sim E^{1/3} \tag{19}$$
$$\text{or} \quad E \sim T^3$$

The development schedule expression is a cubic parabola, and we see deeper meaning in the 1/3 power, cubic expression, as relates volume and length, for example, because of the similarities to technological S-curves described by Richard Foster in Innovation: The Attacker's Advantage[13] to relate development or evolution of Technical Performance measures to Effort. A cubic parabola, $y = ax^{1/3}$, is one good example of such an S-shaped technological curve. This topic deserves further explanation and will be the subject of another paper.

A SOFTWARE DEVELOPMENT EFFORT MODEL

By combining the learning curve term, b, and the software integration term, i, the software development effort exponent becomes (1 + b + i), and the software development effort model proposed by this paper is:

$$MM = PI^{(1+b+i)} \tag{20}$$

where MM = Software development effort, in man-months
      P = Productivity constant, in man-months per thousand lines of source code
      b = Learning curve exponential term = $\ln f / \ln 2$
      I = Integration exponential term = $[\ln (2+f_i)/\ln 2] - 1$

---

[13]Foster, R. N., Innovation: The Attacker's Advantage, Summit Books, New York, 1986.

and f = Learning curve %

   f_i = Relative integration difficulty, in percent. $(0<f_i<1)$

The basic Productivity constant can be derived from the SEL[14] observation that 24 SLOC per day has been an approximately constant average for over a decade. Assuming an 8 hour workday and a 152-hour man-month as in COCOMO, 24 SLOC per day equates to 2.2 man-months per KSLOC. The basic, unmodified value of P is, therefore, taken as 2.2. Many software models have developed effort adjustment factors to account for productivity variations due to program, environment, application, personnel, and other variables. The COCOMO effort adjustment factors, for example, provide an excellent system to modify the average productivity constant for specific projects. Then, $P' = [(2.2)$ times the product of the appropriate Effort Adjustment Factors], can be used to evaluate the software development effort equation:

$$MM = P' \ I^{(1+b+i)} \tag{21}$$

The primary emphasis of this paper is on the exponent. A model has been developed to explain values of the exponent less than or greater than one; that is, upward or downward curves for various projects. The exponent is greater than one and the curve bends upward when integration effects exceed learning effects. On the other hand, when learning effects exceed integration, the exponent is less than one, and the curve bends downward. This model accounts for values of the exponent ranging between 0.78 and 1.50 as indicated from the well-known models listed in Table 1. Values of the learning curve percentage and the integration difficulty percentage corresponding to these well-known models are presented in Table 4. It is noted that nominal values of a 90% learning curve and 55% integration difficulty correspond to the Embedded COCOMO and Jensen exponent of 1.20 or 6/5. Also note that the exponent $(1+b+i)$ is a continuous variable.

The exponent $(1+b+i)$ is useful as such for evaluation of software effort and for estimation of the fraction of total effort required for integration or to be gained by learning curve effects, but it can also be further reduced mathematically as follows:

---

[14]"Proceedings of the Fifteenth Annual Software Engineering Workshop," NASA/GSFC, Greenbelt, MD, November 1990.

TABLE 4

LEARNING CURVE AND INTEGRATION FACTORS
CORRESPONDING TO SOFTWARE MODEL EXPONENTS

Model	Exponent	Value of i(%) if Learning Curve is:			
		80%	85%	90%	95%
Doty Associates (Lowest)	0.781	15%	2%	-	-
Walston and Felix	0.91	35%	21%	9%	-
SEL (Figure 1)	0.92	37%	23%	10%	-
Aron/Wolverton	1.00	50%	35%	22%	11%
COCOMO Organic	1.05	59%	44%	30%	18%
COCOMO Semidetached	1.12	72%	56%	42%	29%
COCOMO Embedded/Jenson SEER	1.20	87%	70%	55%	42%
Doty Associates (Highest)/ Putnam SLIM	1.263	100%	82%	67%	53%
Halstead	1.50	-	-	-	98%

$$1 + b + i \ = \ 1 + \frac{\ln f}{\ln 2} + \left[\frac{\ln (f_i + 2)}{\ln 2} - 1 \right] \tag{22}$$

$$= \ \frac{\ln f + \ln (f_i + 2)}{\ln 2}$$

$$= \ \frac{\ln [f (f_1 + 2)]}{\ln 2}$$

$$= \ \log_2 [f (f_i + 2)]$$

The exponent equals one, for instance, when the learning curve is 80% and
the integration is 50%, since:

$$\log_2 [.8 (2.5)] \ = \ \log_2 [2] \ = \ 1.0 \tag{23}$$

Therefore, the effort estimation curve would bend upward or downward,
with one or the other of f and f_i constant, depending on whether learning
effects were more (and the curve bends down) or integration effects were
more (and the curve bends up).

It is noted that the mathematical derivation of the Jensen[1] effort equation ($E \sim I^{6/5}$) presumes a slope of the productivity relationship of -1/2, and the 1.2 exponent is equivalent to the exponent of the COCOMO Embedded mode equation, whereas Putnam's[8] equations presume a slope of the productivity relationship of -2/3. Therefore, these models make certain assumptions or empirical inferences about the productivity slope which may be related to programmer experience or complexity of application. These inferences lead to a derived value for the exponent. It appears that a productivity slope of -1/2 for a given, difficult application complexity (e.g., in COCOMO Embedded mode) corresponds to a level of programmer experience similar to between 1 and 3 years which corresponds in this model to a learning curve of 90% with an integration difficulty of 55%. Both approaches lead to a 6/5 exponent. As in Putnam[8], a different productivity slope or a different Programmer Experience (and therefore Learning Curve) level lead to a different exponent, be it 1.263 (85% Learning Curve, 82% Integration Difficulty) or 9/7 (85% Learning Curve, 87% Integration Difficulty). We believe the exponent can be estimated from the Learning Curve and the Integration Difficulty, and existing models can be explained as specific cases of this situation. Notice that the Ada COCOMO model utilizes exponent adjustment based upon programmer experience, risk, design thoroughness, and volatility.

At the current state, this model is primarily a conceptual, mathematical model that fits realistic and practical values. The next step is to employ actual project data to calibrate, verify and validate the numerical values of the model.

ESTIMATION OF SOFTWARE INTEGRATION AND CODE REUSE

The software effort estimation model explains the variation in exponents of many well-known models and explains why and under what circumstances the effort estimation curve bends upward and downward. Furthermore, the model can be employed to estimate the amount of software integration effort required to develop a program and also the integration effort required to incorporate previously written, reusable code. Knowledge of the software integration effort estimate is useful in planning, budgeting, and staffing. Reusable code integration effort is a critical factor in the decision to employ reusable code or rewrite.

The software effort estimation model applied to estimate "integration" effort is:

$$MM_{Int} = P' I_{TOTAL}^e - P' \sum_{i=1}^{n} I_i^e \qquad \text{for } e > 1 \tag{24}$$

$$\text{and } I > 1K.$$

In the case of reusable software for a program consisting of I_{new} KSLOC of new code and I_{reuse} KSLOC of reusable code, then:

$$I_{TOTAL} = I_{new} + I_{reuse} \tag{25}$$

and the effort for integration of the reusable code into the software program ($MM_{Int\ reuse}$) is:

$$MM_{TOTAL} = P' (I_{TOTAL})^e - P'(I_{new}^e + I_{reuse}^e) \qquad \text{for } e > 1 \tag{26}$$

$$= P' (I_{new} + I_{reuse})^e - P' (I_{new}^e + I_{reuse}^e)$$

The effort saved by not recreating the reusable code from scratch ($MM_{save\ reuse}$) is:

$$MM_{save\ reuse} = P' I_{reuse}^e \tag{27}$$

So that the total development effort with I_{reuse} KSLOC of reusable code ($MM_{TOTAL + reuse}$) is:

$$MM_{TOTAL + reuse} = [MM_{Intreuse}] + [MM_{writenew}] \tag{28}$$

$$= [P' (I_{new} + I_{reuse})^e - P'(I_{new}^e + I_{reuse}^e)] + P' I_{new}^e$$

$$= P' (I_n + I_r)^e - P' I_n^e - P' I_r^e + P' I_n^e$$

$$= P' (I_n + I_r)^e - P' I_r^e$$

$$= P' (I_{TOTAL})^e - P' I_r^e$$

$$= MM_{TOTAL} - MM_{save\ reuse}$$

If I_{reuse} equals (g)% of I_{new}; that is, if we reuse g% of the amount of new lines of code, then:

$$I_r = g \, I_n \tag{29}$$

and

$$MM_{Int\ reuse} = P' \, I_{TOTAL}^e - P' \, (I_n^e + I_r^e) \tag{30}$$

$$= P' \, (I_n + I_r)^e - P' \, (I_n^e + I_r^e)$$

$$= P' \, (I_n + gI_n)^e - P' \, (I_n^e + g^e I_n^e)$$

$$= P' \, [I_n \, (1 + g)]^e - P'[I_n^e(1 + g^e)]$$

$$= P' \, I_n^e[(1 + g)^e - (1 + g^e)]$$

For example, if the reusable code is 10% of the new code, then the effort to integrate that reusable code into the total program, when the exponent (e = 1+b+i) is 1.20 as in COCOMO Embedded or Jensen, is 5.8% of the new code development effort. A primary use of the integration effort equation occurs when the amount of reusable code exceeds the minimum amount of new code that would be required to do the job.

CONCLUSIONS AND RECOMMENDATIONS

This paper presents and develops a software development effort estimation model that explains how the effort curve bends upward or downward for specific applications, and derives values of the effort estimation exponent that fit other well-known modules. It provides a method to estimate the exponent and the effort for specific software projects based upon learning curves and software integration. It also provides an estimating method for the integration effort associated with reusable code. It is recommended that the next step is to employ actual data to calibrate, verify and validate what is essentially now a mathematical and conceptual model designed to fit practical values.

References

1. R. W. Jensen, "Sensitivity Analysis of the Jensen Software Model," and "A Macro-Level Software Development Cost Estimation Methodology," Fourteenth-Asilomar Conference on Circuits, Systems and Computers, IEEE, New York, 1981.

2. E. A. Nelson, "Management Handbook for the Estimation of Computer Programming Costs" AD-A648750, Systems Development Corp., Oct 31, 1966.

3. J. D. Aron, "Estimating Resources for Large Programming Systems," NATO Science Committee, Rome, Italy, October 1969.

4. R. W. Wolverton, "The Cost of Developing Large-Scale Software, "IEEE Transactions on Computers, pp. 615-636, June 1974.

5. Doty Associates, Inc., "Software Cost Estimation Study, Guidelines for Improved Software Cost Estimating," Volume II, RADC-TR-77-220, August 1977.

6. C. E. Walston and C. P. Felix, "A Method of Programming Measurement and Estimation," IBM System Journal, 16, 1, 1977.

7. P. V. Norden, "Project Life Cycle Modeling: Background and Application of Life Cycle Curves," Software Life Cycle Management Workshop, USACSC, Airlie, VA, August 1977.

8. L. H. Putnam, "Example of an Early Sizing, Cost, and Schedule Estimate for an Application Software System," COMPSAC 1978, November 13-16, 1978.

9. B. W. Boehm, Software Engineering Economics, Prentice-Hall, Inc., Englewood Cliffs, NJ, 1981.

10. H. R. Barton, Jr., "Predicting Guaranty Support Using Learning Curves," Proceedings of the Annual Reliability and Maintainability Symposium, IEEE, 1985.

11. J. T. Duane, "Learning Curve Approach to Reliability Monitoring," IEEE Transactions on Aerospace, Vol. 2, No. 2, April 1964.

12. A. J. Albrecht and J. E. Gaffney, Jr., "Software Function, Lines of Code, and Development Effort Prediction: A Software Science Validation," IEEE Transactions on Software Engineering, November 1983.

13. R. N. Foster, Innovation: The Attacker's Advantage, Summit Books, New York, 1986.

14. "Proceedings of the Fifteenth Annual Software Engineering Workshop," NASA/GSFC, Greenbelt, MD, November 1990.

Software Cost Estimating Models: A Calibration, Validation, and Comparison

Gerald L. Ourada and Daniel V. Ferens

Air Force Institute of Technology, Wright-Patterson Air Force Base, OH

ABSTRACT

This study was a calibration, validation and comparison of four software effort estimation models. The four models evaluated were REVIC, SASET, SEER, and COSTMODL. A historical database was obtained from Space Systems Division, in Los Angeles, and used as the input data. Two software environments were selected, one used to calibrate and validate the models, and the other to show the performance of the models outside their environment of calibration.

REVIC and COSTMODL are COCOMO derivatives and were calibrated using Dr. Boehm's procedure. SASET and SEER were found to be uncalibratable for this effort. Accuracy of all the models was significantly low; none of the models performed as expected. REVIC and COSTMODL actually performed better against the comparison data than the data from the calibration. SASET and SEER were very inconsistent across both environments.

1. INTRODUCTION

Because of the tremendous growth in computers and software over the last twenty years, the ability to accurately predict software life cycle cost is critical to Department of Defense (DoD) and commercial organizations. To predict these costs, numerous software cost estimation models have been developed; however, the accuracy, and even the usability of these models for DoD and other software projects is questionable. The available models have not received a significant amount of rigorous calibration and testing from a solid historical data base [7: 558-567]. Furthermore, the requisite data collection and model analysis has not been performed during software acquisition projects to demonstrate model accuracy.

This paper primarily addresses a study performed at Air Force Institute of Technology (AFIT) as a thesis effort from September, 1990 to July, 1991 to determine whether some existing cost models can be calibrated and validated on DoD projects to establish their relative accuracy. First, a summary of past efforts in model validation is presented as background information. Next, the AFIT study is described in detail with respect to the calibration and validation of selected cost models. Finally, some recommendations regarding this study and software cost estimation in general are presented.

2. PAST VALIDATION EFFORTS

There have been several efforts in the past to qualitatively and quantitatively assess software cost models. A summary of some of these efforts is presented here to demonstrate the current status of cost model studies.

Qualitative Studies. The primary purpose of these studies is to determine the suitability, or usability of software cost models for particular projects or organizations. They do not assess probable accuracy or other quantitative factors; however, they are useful in assessing which model or models may best satisfy a user's needs. Some examples of past qualitative studies are now discussed.

A comprehensive study for DoD was performed by the Institute for Defense Analysis [1] using five criteria to evaluate seven software cost models. The five criteria were: (1) Assistance in making investment decisions early in the life cycle, (2) Assistance in validating contractor proposals, (3) Support for day-to-day project activities, (4) Assistance in predicting software maintenance, and (5) Support to identify major cost drivers and productivity improvements. Of the seven models studied, it was found that the PRICE-S model, the Software Life Cycle Management (SLIM) model, and the Jensen-3 model (a forerunner to the current SEER and SYSTEM-4 models discussed later) were most useful for the first three criteria; the Software Productivity, Quality, and Reliability - 20 (SPQR/20) and SLIM were most useful for the fourth criterion, and the SoftCost-R model was most useful for the fifth criterion. A version of the Constructive Cost Model (COCOMO) used in this study did not excel in any of the criteria, but was cited for its low cost and visibility into the model algorithms.

Several qualitative studies have been performed for software maintenance, or support cost models; these studies can be especially useful since, as of this time, no software support cost models have been quantitatively validated for any applications. A study performed by one of the authors [8: 2-12] concluded that, of eight models studied, SEER, PRICE-S, and the DoD-owned SASET models considered support costs most thoroughly; however, certain organizations have preferred less thorough models because of cost, ease-of-use, or other considerations. Another study by personnel from Headquarters, Air Force Logistics Command [3: 13-25] analyzed the capabilities of COCOMO, PRICE-S, SASET, and SLIM for software support cost estimation. This study did not advocate the use of a particular model, but presented some of the problems associated with software support cost estimation in general.

Quantitative Studies. One of the earliest comprehensive model validation studies wan performed by Robert Thibodeau [14], which investigated nine software cost models including early versions of PRICE-S and SLIM. The study compared the estimates of the models to actual values for three data bases; an Air Force data base for information systems software containing seventeen values, a military ground systems software data base of seventeen data points, and a commercial software data base containing eleven values. The study showed that SLIM, when calibrated, averaged within 25% of actual values for commercial and information systems software; and PRICE-S, when calibrated, averaged within 30% of actual values for military ground systems software. It was also discovered that when both models were not calibrated, their accuracies were about five times worse. Although the Thibodeau study did not address recent data or models (except PRICE-S and SLIM), it did demonstrate the necessity for model calibration, and that different models were more accurate for different environments.

A more recent study was performed by Illinois Institute of Technology [9] with eight Ada language projects and six cost models: SYSTEM-3 (a forerunner to SYSTEM-4 and SEER), PRICE-S, SASET, SPQR/20, and the Ada versions of COCOMO and SoftCost-R. The eight Ada projects were divided into three sub-categories: object-oriented versus structured design; government versus commercial contracts, and command and control versus tools/environment applications. The estimates of the models (which were not calibrated to the data base) were compared to actual results, and the models were rank-ordered based on how many estimates were within 30% of actual values. SoftCost-Ada and SASET scored highest on overall accuracy; however, models varied in results for sub-categories. For example, SASET and SPQR/20 scored highest for command and control applications while SoftCost-Ada scored highest for tool/environment applications. The models were also evaluated for consistency of estimates to within 30% after the mean for the model's estimate was applied. Here, PRICE-S and SYSTEM-3 scored highest; which showed that calibration may enhance the accuracy of these models. The results of this study are consistent with Thibodeau's study in that different models performed better for different (Ada) applications, and calibration can improve model results.

Several other quantitative studies have been done to assess model accuracy; however, for the DoD environment, the studies have not generally employed a significantly large data base, used models calibrated to the data studied, or attempted to validate the models on additional data sets. The AFIT study attempted to address these issues and others for the DoD environment. It should be noted that this study is restricted to development cost estimation and does not address support costs, development schedules, etc.

3. THE AFIT STUDY

Research Objectives. This research addresses the following set of questions:

1. Given a credible set of actual DoD data, can the chosen models be calibrated?
2. Given a calibrated model, with another set of actual data from the same environment, can the models be validated?
3. Given a validated model, if another independent data set from another software environment is used, are the estimates still accurate?
4. Is a calibration and validation of a model accurate for only specific areas of application?

Scope of Research. Since effort estimation models can be expensive, this research was limited to models existing at the Air Force Institute of Technology (AFIT) or available from other government sources. Currently there are eight such models:

1. REVIC (REVised version of Intermediate COCOMO);
2. COCOMO (COnstructive COst MOdel);

3. PRICE-S (Programmed Review of Information for Costing and Evaluation Software);
4. SEER (System Evaluation and Estimation of Resources);
5. SASET (Software Architecture, Sizing and Estimating Tool);
6. System-4;
7. Checkpoint/SPQR-20;
8. COSTMODL (COST MODeL).

Time constraints restricted this research to four models. The following are the four selection criteria used to guide the selection of models to study:

1. Use within DoD or NASA;
2. Ease of understanding and analyzing the input and the output;
3. Availability of model documentation;
4. Cost to use the models for this research effort.

The above criteria were derived from personal experience in project management within DoD and the potential for cost to impact the research effort. Only those models that are relatively easy to use and understand will be used by any project team. Also if the model already belongs to the government, then there exists a greater chance of the model being used due to less cost to the potential user.

The four models selected were, REVIC, SASET, SEER, and COSTMODL. For each of these models, either DoD or NASA has a license to use or is the owner of the model.

Methodology. This research was conducted in three parts: model calibration, validation, and comparison. During calibration the model parameters were adjusted to give an accurate output with known inputs. One-half of the database, selected at random, was used as input data. The model parameters were then adjusted mathematically to give an output as close as possible to the actual output contained in the data base. The particular calibration technique is dependent upon the particular model under evaluation; the technique suggested in the model users guide was used. Once the model was calibrated, the model was analyzed with the calibration data set to examine the model for accuracy against the calibration data.

During validation, the second half of the database was used. In this phase the input data is used, but the model parameters were not changed. The objective is to examine the statistical consistency when comparing the known output to the estimated output [5: 175-176]. The validation data set was entered in the models, and the results analyzed for accuracy. This validation should show that the model is an accurate predictor of effort in the environment of the calibration.

The third part of the research was a run of the independent data set through the models to examine the validity of the model outside its calibrated environment. The effort estimations were then analyzed for accuracy against the actual effort. The accuracy analysis should show that outside the environment of calibration, the models do not predict well, i.e. a model calibrated to a manned space environment should not give accurate estimates when used to estimate the effort necessary to develop a word processing application program.

To test the accuracy of the models, several statistical tests are used. The first tests are the coefficient of multiple determination (COMD or R^2) and the magnitude and mean magnitude of relative error. For the coefficient of multiple determination, Equation 1, E_{act} is the actual value from the database, E_{est} is the estimate from the model, and E_{mean} is the mean of the estimated values, Equation 2. The COMD indicates the extent to which E_{act} and E_{est} are linearly related. The closer the value of COMD is to 1.0, the better. (It is possible to get negative values for COMD if the error is large enough. The negative values appear when the difference between the actual effort and the estimate is extremely large.) A high value for COMD suggests that either a large percentage of variance is accounted for, or that the inclusion of additional independent variables in the model is not likely to improve the model estimating ability significantly. For the model to be considered calibrated, values above 0.90 are expected [5: 148-176].

$$R^2 = 1 - \frac{\sum\limits_{i=1}^{n} (E_{act_i} - E_{est_i})^2}{\sum\limits_{i=1}^{n} (E_{act_i} - E_{mean})^2} \tag{1}$$

$$E_{mean} = \frac{1}{n} * \sum\limits_{i=1}^{n} E_{est_i} \tag{2}$$

The equation for magnitude of relative error (MRE) is Equation 3, and for mean magnitude of relative error (MMRE), Equation 4. A small value of MRE indicates that the model is predicting accurately. The key parameter however, is MMRE. For the model to be acceptable, MMRE should be less than or equal to 0.25. The use of MRE and MMRE relieve the concerns of positive and negative errors canceling each other and giving a false indication of model accuracy [5: 148-176].

$$MRE = \left| \frac{E_{act} - E_{est}}{E_{act}} \right| \tag{3}$$

$$MMRE = \frac{1}{n} * \sum_{i=1}^{n} MRE_i \tag{4}$$

Errors using the MRE and MMRE tests can be of two types: underestimates, where $E_{est} < E_{act}$; and overestimates, where $E_{est} > E_{act}$. Both errors can have serious impacts on estimate interpretation. Large underestimates can cause projects to be understaffed and, as deadlines approach, project managers will be tempted to add new staff members, resulting in a phenomenon known as Brooks's law: "Adding manpower to a late software project makes it later" [4: 25]. Large overestimates can also be costly, staff members become less productive (Parkinson's law: "Work expands to fill the time available for its completion") or add "gold-plating" that is not required by the user [10: 420].

The second set of statistical tests are the root mean square error (RMS), Equation 5, and the relative root mean square error (RRMS), Equation 6. The smaller the value of RMS the better is the estimation model. For RRMS, an acceptable model will give a value of RRMS < 0.25 [5: 175].

$$RMS = \sqrt{\frac{1}{n} \sum_{n=1}^{n} (E_{act} - E_{est})^2} \tag{5}$$

$$RRMS = \frac{RMS}{\frac{1}{n} \sum_{n=1}^{n} E_{act}} \tag{6}$$

The third statistical test used is the prediction level test, Equation 7, where k is the number of projects in a set of n projects whose MRE is less than or equal to a percentage l.

$$PRED(l) = \frac{k}{n} \qquad\qquad (7)$$

For example, if PRED (0.25) = 0.83, then 83% of the predicted values fall within 25% of their actual values. To establish the model accuracy, 75% of the predictions must fall within 25% or the actual values, or PRED (0.25) >= 0.75 [5: 173].

Analysis and Findings. For this research effort, the November 1990 version of a data base collected by SSD/ACC (Space Systems Division, Los Angeles AFB) was used. The updated database will eventually contain over 512 data points with a large amount of information for each point. The November 1990 version had enough data points, 150, that the methodology discussed could be used. The actual data in this database cannot be published due to the proprietary nature of the information.

The SSD database was searched for at least 20 data points which could be used for the calibration and validation attempts. Twenty-eight data points were found that; had the same development environment (Military Ground Systems), had data for the actual development effort, had no reused code, and were similar sized projects. Having no reused code was a necessary requirement since the database does not include any information about the distribution of reused code, i.e. the amount of redesign, recode, etc., to determine the estimated source lines-of-code (SLOC) necessary for the model inputs. The selected project size ranged from 4.1K SLOC to 252K SLOC. Fourteen of the data points were used for the calibration effort and the other 14 for the validation effort. The selection of which 14 went to which effort was made by alternating the selection of the projects; the first went to the calibration effort, the second went to the validation effort, the third to calibration, etc.

For the comparison part of this research, 10 projects were found in the SSD database which fit all of the above criteria except for the development environment. The development environment selected was Unmanned Space Systems since data was available and this environment is different than Military Ground Systems.

REVIC. Since REVIC is a COCOMO derived estimation model [11], the technique described by Dr. Boehm [2: 524-530] was used to perform the calibration. Dr. Boehm recommends at least 10 data points should be available for a coefficient and exponent calibration. Since 14 data points were available, the coefficient and exponent calibration was performed initially. However, since the number of data points was not large, this researcher decided to perform a coefficient only calibration also and compare the two calibrations. The semi-detached mode (Equation 8) of REVIC was used for the calibration and validation since the description of the projects selected from the SSD database for calibration and validation fit the description of Dr. Boehm's semi-detached mode, where MM is the output in man-months, kDSI is the source lines of code in thousands, and $\prod$ is the product of the costing parameters [2: 74-80, 116-117].

$$MM=3.0x(kDSI)^{1.12}\prod \tag{8}$$

The embedded mode (Equation 9) was used in the comparison analysis for the coefficient only calibration since these data points match the description of Dr. Boehm's embedded mode description [2: 74-80, 116-117].

$$MM=2.8x(kDSI)^{1.20}\prod \tag{9}$$

REVIC Calibration. The adjustment of the input values will give the calibrated coefficient and exponent or coefficient only values for this particular data set. For the coefficient and exponent calibration, the calibrated output values were 2.4531 and 1.2457 respectively. For the coefficient only calibration, the REVIC calibrated exponent of 1.20 was used. The calibrated coefficient was found to be 3.724.

These new coefficients and exponents were then put back into the estimation equations to look at prediction accuracies of the model for the data used for calibration. Table 1 shows the results of the accuracy analysis.

The interesting item of note here is that, for all the parameters, the coefficient only calibration appears to be more accurate than that of the coefficient and exponent. This may be explained by the fact that the exponent calibration is very sensitive to small variations in project data [2: 524-529]. With a larger calibration data set the accuracy of the coefficient and exponent calibration may be better.

The other interesting item of note is the general accuracy of the calibrated model. Even against the calibration data, the model is not inherently accurate. R^2 should be greater than 0.90, MMRE and RRMS should be less than 0.25, RMS should be small (approaching 0), and PRED(0.25) should be greater than 75%. The coefficient only results approach acceptability as defined by Conte [5: 150-176], but are nowhere near what should be expected of a model when tested against its calibration data.

REVIC Validation. The results of the accuracy analysis for validation are shown in Table 2. Again, analysis of this table shows the coefficient calibration to be more accurate than the coefficient and exponent calibration. However, in this case both calibrations were able to predict four of the 14 validation projects to within 25% of their actuals. The differences in R^2, MMRE and RRMS show that the coefficient only calibration was more accurate, but none of the values are near what would be expected to say this model is validated to this environment.

REVIC Comparison. The embedded mode was used for the coefficient only analysis with the new calibrated coefficient used. The results of the comparison accuracy analysis are shown in Table 3.

These results almost show this research effort to be futile, at least for the REVIC estimation model. The results show that both calibration efforts are fairly accurate with this set of data. Even though the PRED was low, the other parameters are all very close to, if not, acceptable values. The R^2, MMRE, and RRMS show better results for the coefficient and exponent calibration, but the PRED and MMRE are much better for the coefficient only calibration. These results make this researcher question this model, using either the coefficient only or the coefficient and exponent calibration, as a valid effort estimation tool for any software manager. The model is too good at estimating outside the environment of calibration and not good at all inside the environment.

SASET. The research effort using the SASET estimation model was very frustrating. As this author reviewed the SASET model and User's Guide [13], the ability to calibrate the model was found to be virtually impossible. Since the mathematical equations published with the users guide are virtually impossible to understand, for the "average" user, and a calibration mode is not available as part of the computerized version of the model, this author could not figure out how to calibrate the model to a particular data set. The only way found to perform a calibration was to go into the calibration file of the computerized model and change the actual values of several hundred different parameters. Without the knowledge of what each of these parameters actually does within the model, any changes would be pure guesswork. Again, the User's Guide was of no help. This model has an unpublished saying that accompanies it, "There are no casual users of SASET." This saying seems very true, because an informal survey of normal users of effort estimation models revealed that they do not have the time, and sometimes not the mathematical abilities, to figure out the intricacies of this model.

Because of the above factors, a calibration of SASET was not accomplished. However, this research effort used SASET with its delivered calibration file and the 28 calibration and validation and 10 comparison data points were input to the model to test the model with its delivered calibration.

SASET Calibration/Validation. Table 4 shows the accuracy results for the calibration, validation, and comparison data sets. As can be seen from the data, the existing calibration of SASET is very poor for this data set. The estimates were all greater than the actuals, with estimates from 2 to 16 times the actual values given as outputs from the model. The negative values of R^2 are a result of the large differences between the actual effort and the estimate from the model.

SASET Comparison. The comparison data was analyzed with the SASET model to see if another environment was any better with the delivered calibration. As can be seen by the data in Table 4, the comparison data set also shows a very poor calibration for the data set. All of the estimates were greater than the actual efforts, nine of the ten data points were estimated between two and three times the actuals. This does at least show some consistently high estimation.

For the SASET model the computerized version is delivered with one specific calibration. For the layman software effort estimator, this model has very questionable usability in its current form.

SEER. SEER was also found to be a problem for this research effort; however, this issue was not because of the usability (or unusability) of the model. The SEER model is calibratable, but only if the data set is properly annotated [12]. The model has a parameter called "effective technology rating" which is used to calibrate the model to a particular environment or data set. To perform the evaluation of the effective technology parameter with a historical data set, the actual effort for the Full Scale Implementation phase (a SEER term) must be known. This phase does not include requirements analysis, or system integration and testing. The database that was used for this effort includes the necessary data, but not to the detail necessary to perform the calibration; i.e. the actual effort is known, but the effort during Full Scale Implementation is not.

SEER Calibration/Validation. The 28 data points of the calibration and validation data set were ran through the model to test for model accuracy with this particular environment. Table 5 shows the results of this accuracy analysis.

The estimates from the model ranged from 25% of the actual to 11 times the actual effort. Most of the estimates were in the range of 2-5 times the actual. The results shown in Table 5 again show the need to calibrate a model to a particular environment. R^2 is negative due to the large differences between the actual effort and the estimated effort .

SEER Comparison. The comparison data was also ran through the model. The results of the accuracy analysis are also shown in Table 5. These results are some what better than those for the calibration and validation, but again this model, as calibrated, should not be used in these environments. The estimates for this data set were all greater than the actual, ranging from very near the actual to three times the actual value.

The results of the accuracy analysis, especially the comparison data, lead this researcher to conclude that the SEER model may have some use if a proper calibration can be accomplished; but this will require a historical database that has the necessary effort information in each phase of the development life-cycle.

COSTMODL. The first review of COSTMODL [6] revealed several differences between it, COCOMO, and REVIC. For this reason it was selected as a model to be evaluated. However, once a good database was found, the only implementation of the model that was still valid (i.e. a non-Ada version) was that of the original COCOMO, adjusted to account for the Requirements Analysis and Operational Test and Evaluation phases. The procedure explained by Dr. Boehm [2: 524-530] was used to perform the calibration.

COSTMODL Calibration. Since REVIC was analyzed for both the coefficient only and coefficient and exponent, COSTMODL was also. The derived coefficient only coefficient value was 4.255. The values for the coefficient and exponent analysis were 3.35 and 1.22 for the coefficient and exponent respectively. These values were then used to replace the original coefficients and exponents in the model, and the model was analyzed for accuracy against the calibration data set. Table 6 shows these results.

These values are very similar to the accuracies shown with REVIC. This model is calibratable, but it still leaves a lot to be desired in the accuracy area. The coefficient only calibration appears to perform somewhat better against the calibration data set, but the performance increase is very small.

COSTMODL Validation. The validation data set was ran and again analyzed for accuracy. The results are shown in Table 7. Again, the coefficient only calibration appears to be a better estimator of the actual effort. The results of this accuracy analysis show a questionable estimation model for the COSTMODL effort estimation, and the COCOMO baseline equations. These results are nowhere near what are necessary for a usable model within DoD.

COSTMODL Comparison. The comparison data was used to see the effect of using a estimation model outside its calibrated environment. The accuracy analysis is shown in Table 8.

Analysis of this data shows that this is a good calibration for this data set. This is not supposed to happen; a model should not work this well outside its calibrated environment. This researcher does not understand why this model predicts well outside its environment of calibration.

The coefficient only comparison analysis uses the embedded mode of Intermediate COCOMO, the same as with the REVIC comparison analysis.

Conclusions. This research proved to be very enlightening. The two models that could be calibrated, REVIC and COSTMODL, could not predict the actuals against either the calibration data or validation data to any level of accuracy or consistency. Surprisingly, both of these models were relatively good at predicting the comparison data, data which was completely outside the environment of calibration. For the two models which were not calibrated, SASET and SEER, it was shown that calibration is necessary, but may not be sufficient to make either of these models usable.

One interesting item that was found: During the initial attempts at calibrating REVIC and COSTMODL, one data point was used which had a significantly larger amount of code than any of the others (over 700 KSLOC vs. less than 250 KSLOC). This one data point was found to drive any attempt at calibration. The amount of code is one of the key terms used in the calibration technique for COCOMO and derivatives [2: 524-530]. This number is squared in several places as part of the calibration, and when one of the data points is much larger than the others, this squaring creates an extremely large number that can be magnitudes larger than those for the other data points. When these squared values are then summed, this one data point can drive the value of the sum. Therefore, this data point was removed from the calibration database.

REVIC proved to be a fairly easy model to learn and use. The calibration was not difficult and did produce an increased ability to estimate effort compared to the original calibration. However, the accuracy of this model is questionable based upon the results found in this research effort. This researcher found it interesting that the coefficient only calibration was actually more accurate than the coefficient and exponent calibration. This can probably be explained by the sensitivity of the exponent, but no way to test this is known by this researcher.

SASET proved to be the most difficult model to learn and use. The User's Guide is very unclear, and the model is not easy to learn and use just by running the computerized program. The calibration for this model will probably prove to be virtually impossible for any user other than one of the model developers. This alone makes this model very difficult to use for any DoD acquisition program office since calibration is apparently needed. The model has many nice features and is very flexible in allowing risk analysis and trade-off analysis; but, if the model cannot be calibrated to the working environment, it probably cannot be used as an accurate predictor in a program office.

SEER was a fairly easy model to learn and use. The User's Guide is very well written and is easy to follow, once the template structure is learned. This model is relatively easy to calibrate if the historical data can be put into the necessary format. The inaccuracies found with the estimation analysis proved that SEER also needs to be calibrated to the operating environment. This should be done soon, since the Air Force will have a site license for this model beginning fiscal year 1992.

COSTMODL turned out to be very similar to REVIC. The model was very easy to learn, understand and use. Here the coefficient only calibration also seemed to work better than the coefficient and exponent calibration. This model proved to be calibratable, but again the poor accuracy results make it a questionable resource for any program manager.

4. RECOMMENDATIONS

The results of this study should not discourage further efforts to calibrate and validate software cost estimating models for DoD and other software projects. Since past studies have demonstrated that specific models appear to be accurate for particular environments, it is probable that one or more models can be shown accurate for DoD programs. What is needed is continued efforts to establish larger and more complete historical data bases for model calibration and validation. Additionally, a matrix of inputs, outputs, phases considered, and other model-unique factors is essential for direct comparison of model results when two or more models are being compared. Furthermore, thorough education of software cost analysts is needed to insure the individuals performing studies (as well as software cost estimates in general) are competent. Finally, a complete data base for software support cost estimation should be developed; none exist at the current time.

Table 1   REVIC Calibration Accuracy Results

	Coefficient and Exponent	Coefficient Only
R^2	0.776	0.892
MMRE	0.3733	0.334
RMS	119.1416	82.641
RRMS	0.3192	0.221
PRED (0.25)	42%	57%

Table 2   REVIC Validation Accuracy Results

	Coefficient and Exponent	Coefficient Only
R^2	0.1713	0.6583
MMRE	0.7811	0.6491
RMS	375.190	211.020
RRMS	0.8560	0.4815
PRED (0.25)	28.5%	28.5%

Table 3   REVIC Comparison Accuracy Results

	Coefficient and Exponent	Coefficient Only
R^2	0.9081	0.8381
MMRE	0.2201	0.1767
RMS	66.161	87.844
RRMS	0.2069	0.2748
PRED (0.25)	30%	70%

Table 4   SASET Accuracy Results

	Calibration/ Validation	Comparison
R^2	-0.7333	-0.3272
MMRE	5.9492	1.0985
RMS	1836.4	527.6
RRMS	4.5097	1.6503
PRED (0.25)	3.5%	0%

Table 5   SEER Accuracy Results

	Calibration/ Validation	Comparison
R^2	-1.0047	-0.2529
MMRE	3.5556	0.5586
RMS	1504.9	380.6
RRMS	3.6955	1.1905
PRED (0.25)	10.7%	20%

Table 6   COSTMODL Calibration Accuracy Results

	Coefficient and Exponent	Coefficient Only
R^2	0.5251	0.760
MMRE	0.4603	0.396
RMS	175.57	124.27
RRMS	0.4703	0.333
PRED (0.25)	29%	35.7%

Table 7   COSTMODL Validation Accuracy Results

	Coefficient and Exponent	Coefficient Only
R^2	0.1120	0.6353
MMRE	0.7863	0.5765
RMS	411.516	220.667
RRMS	0.9389	0.5035
PRED (0.25)	21.4%	21.4%

Table 8   COSTMODL Comparison Accuracy Results

	Coefficient and Exponent	Coefficient Only
R^2	0.8661	0.8369
MMRE	0.2003	0.1751
RMS	79.454	87.94
RRMS	0.2485	0.2751
PRED (0.25)	70%	60%

REFERENCES

1. Bailey, Elizabeth K., et al, A Descriptive Evaluation of Automated Software Cost Estimation Models (IDA Paper P-1979), Washington, DC, Institute for Defense Analysis: October, 1986.

2. Boehm, Barry. Software Engineering Economics, Englewood Cliffs NJ: Prentice-Hall, 1981.

3. Boulware, Gary W., et al, "Maintenance Software Model Assumptions Versus Budgeting Realities", National Estimator, Spring, 1991, pp 13-25.

4. Brooks, Frederick P. Jr. The Mythical Man-Month, Menlo Park CA: Addison-Wesley, 1982.

5. Conte, Samuel D., and others. Software Engineering Method and Metrics. Menlo Park CA: Benjamin/Cummings, 1986.

6. COSTMODL User's Guide, NASA Johnson Space Center, Houston TX, version 5.2, Jan 1991.

7. Cuelenaere, A.M.E., and others. "Calibrating a Software Cost Estimation Model: Why & How," Information and Software Technology, 29: 558-567 (10 Dec 1987).

8. Ferens, Daniel V., "Evaluation of Eight Software Support Cost Models", National Estimator, Spring, 1991, pp 2-12.

9. Illinois Institute of Technology (IIT) Research Institute, Test Case Study: Estimating the Cost of Ada Software Development, Lanham, MD, IIT: April, 1989.

10. Kemerer, Chris F. "An Empirical Validation of Software Cost Estimation Models," Communications of the ACM, 30: 416-429 (May 1987).

11. Kile, Raymond L. REVIC Software Cost Estimating Model User's Manual, version 9.0, Feb 1991.

12. SEER User's Manual, Galorath Associates Inc., Marina Del Rey CA, Aug 1989.

13. Silver, Dr. Aaron, and others. SASET Users Guide, Publication R-0420-88-2, Naval Center for Cost Analysis, Department of the Navy, Washington DC, Feb 1990.

14. Thibodeau, Robert, An Evaluation of Software Cost Estimating Models, Huntsville, AL, General Research Corporation: 1981.

Cost Estimating for Automated Information Systems

William Richardson
The Analytic Sciences Corporation, 2555 University Boulevard, Fairborn, OH 45324

The establishment of the Major Automated Information System Review Council (MAISRC) has had, and will continue to have significant impacts on the cost estimating community. The three major sections of this paper examine the MAISRC review process in general with emphasis on the role of the cost estimator.

The first section traces the evolution of, and provides introductory information on the MAISRC; its purpose, why it was established, the Council principals, and the MAISRC thresholds are included. The role of OASD(PA&E) is also explained.

The automated information system (AIS) acquisition process, is discussed next. The six AIS life cycle phases are defined along with the purpose of, and activities associated with, each phase. The four MAISRC cost products (i.e., POM/BES, program manager's estimate, independent cost estimate, and the economic analysis) and the milestone reviews for which each is required are identified. Also included is a discussion of the time required to complete and present the MAISRC cost products.

The paper concludes with a discussion of some of the cost estimating differences between weapons system and AIS cost estimating.

MAISRC EVOLUTION

History — The MAISRC was initially established in 1978 as the DoD review and oversight authority for major AIS programs. At that time, AIS programs were relatively few in number, of low complexity, and inexpensive. As a result they were relatively easy to manage and the MAISRC was able to keep a low profile. Since then, things have changed rapidly.

The decade of the 1980's was one of rapid expansion and technological change for the automatic data processing industry. Consequently, DoD's automated information systems became increasingly complex. Cost increased along with complexity as did the number of AIS's. Operational commands with little or no experience in managing large, complex, and costly systems suddenly found themselves in the "acquisition" business responsible for managing and acquiring large systems that often rivaled weapon systems in terms of technical complexity and cost. Primarily because of the cost growth, Congressional interest was aroused in the mid–1980's and a low profile was no longer possible for the MAISRC.

Congress found that (1) existing MAISRC procedures were okay but not always followed, and (2) program fragmentation was widespread. Program fragmentation is the practice of dividing a large program into several smaller pieces and managing each piece as a separate program. Small programs have at least two advantages over large ones. First, the review and oversight process is less stringent and, second, small programs are less likely to be questioned during Congressional budget appropriations hearings. Consequently, Congress directed DoD to (1) eliminate program fragmentation, and (2) employ a "systems" approach for AIS management similar to that used to manage weapon system programs.

Systems Approach — As illustrated in Fig. 1, the systems approach recognizes that an AIS is more than just the prime mission equipment (i.e., hardware and software). It also includes all of the peripheral elements such as maintenance, support equipment, buildings and other facilities, people, training, supplies, user manuals, and other technical data. In short the system includes everything necessary to operate and maintain it as a mission ready unit (Ref. 1).

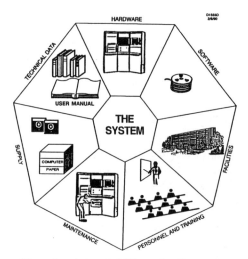

Figure 1   Automated Information Systems

MAISRC Thresholds — The system is considered major and, therefore, subject to the MAISRC review process, if it meets one or more of the following three criteria (Ref. 2):

- Total acquisition cost is anticipated to exceed $100 million. Total acquisition cost includes all costs from the beginning of the Need Justification phase to the end of the Deployment phase.

- The estimated acquisition costs are greater than $25 million in any one year.

- It is designated by OSD as a special interest program.

The Role of OASD(PA&E) — In implementing the Congressional direction for a systems approach, DoD provided an expanded role for the Assistant Secretary of Defense for Program Analysis and Evaluation (OASD(PA&E)). The responsibilities of OASD(PA&E) are carried out by the Cost/Benefit Review Group (CBRG). The CBRG acts as the principal advisory body to the MAISRC on matters related to AIS cost and cost related benefits. Their role is similar to that of the Cost Analysis Improvement Group for weapon system reviews. The primary responsibilities of the CBRG are to (Ref. 2):

- Provide an independent review and validation of AIS life cycle costs and benefits to ensure that they are realistic, complete, and consistent.

- Establish criteria and procedures concerning the preparation, presentation, documentation, estimation, and validation of AIS life cycle costs and cost related benefits.

- Prepare the MAISRC report containing the CBRG's findings, recommendations, and any cost or benefit estimates conducted by the CBRG.

Typically, cost and benefit estimates are segregated and reviewed as separate entities by two different analysts. That is, cost estimates are reviewed by a cost analyst and benefit estimates by a benefits analyst.

The primary goal of the CBRG is to improve the overall AIS management process. Their participation in the review of cost and benefit estimates has contributed to that goal in at least three ways (Ref. 3).

- First, program managers and technical personnel are now forced to define a more complete system description early in the life cycle. As early as Milestone I cost and benefit estimates are subject to a very rigorous, intense review process and must be able to withstand a high degree of scrutiny. This requires a system description so complete that almost everything but the brand name should be known. This does not mean that the system description is "set in concrete" at Milestone I and will never change. Changes are inevitable as the system matures and user requirements become better understood.

- Second, all program office personnel, but especially cost and benefit estimators, are forced to have a more thorough understanding of the system requirements and how they impact expected costs and benefits. A reasonably good knowledge of how the system works is necessary to conduct credible initial cost and benefit estimates. It is even more important when requirements change and it becomes necessary to update the initial estimates. Without a thorough understanding of the initial system requirements, it is nearly impossible to estimate the impacts that changes to those requirements will have on costs and benefits. Simply put, "If you don't understand it, you can't estimate it."

- A third area deals with the enhancement of the AIS estimating function. Clearly, CBRG participation has brought a greater degree of discipline and structure to the AIS estimating process. Just knowing that an estimate is going to be scrutinized by senior, knowledgeable analysts forces the estimator to be more careful and thorough.

A final point of emphasis on the PA&E role is the importance of getting the CBRG involved early in the estimating process. This is one of the CBRG's objectives, but the cost estimator should take the initiative to find out who the action officer(s) will be and make the initial contact. Explain any unique characteristics or issues associated with the program. The chances of resolving controversial issues are greatly improved if all parties are aware of them while there is still time to work the problems. The action officer may even be able to suggest acceptable ways of handling them. The entire estimating and review process will go smoother if the estimator knows the reviewer's expectations.

MAISRC Principals — In addition to OASD(PA&E), the other MAISRC Principals are as follows (Ref. 4):

- The Assistant Secretary of Defense (Comptroller) chairs the MAISRC
- Assistant Secretary of Defense (Command, Control, Communications, and Intelligence)
- The OSD System Sponsor
- The Assistant Secretary of the Service or Agency sponsoring the AIS
- Director, Operational Test and Evaluation
- Other OSD principals identified by the MAISRC.

THE MAISRC PROCESS CONTINUES THROUGHOUT THE LIFE CYCLE

Life Cycle Phases — The "cradle–to–grave" concept applied to weapon systems also applies to automated information systems. The AIS life cycle consists of six phases and covers the period from the program's initial inception to its disposal at the end of its useful life. The six (sometimes overlapping) phases and their relationship to cost are depicted in Fig. 2. Like weapon systems each phase, except for the Disposal phase, is more expensive than the last. Also like weapon systems (though not shown in Fig. 2), the majority of the life cycle costs are designed into the system very early in the life cycle. The MAISRC review and approval process continues through the entire life of the AIS.

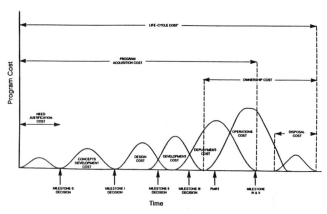

Figure 2   Automated Information System Life Cycle

MAISRC Process (Ref. 4) — Similar in concept and structure to the Defense Acquisition Board process used for weapon systems, major AIS programs are formally reviewed by the MAISRC at key milestones throughout the life cycle. As the senior DoD management oversight committee for automated information systems, MAISRC approval is required before the program is allowed to proceed to the next phase. The MAISRC process, therefore, is a combination of reviews and approvals. As illustrated in Fig. 3, each phase consists of activities which lead to a milestone approval action at the end of the phase. For example, the activities during the Concepts Development Phase are the evaluation of potential alternatives and the selection of the preferred alternative. Once the preferred alternative is formally reviewed and approved by the MAISRC at the Milestone I review, the Design Phase can begin. Milestone reviews occur at the end of each of the first four phases (Milestones 0, I, II, and III) and twice during the Operations Phase.

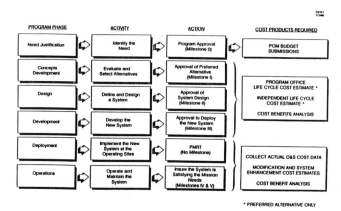

Figure 3   AIS Acquisition Process

During the Need Justification Phase, the requirement for a new or modified AIS is identified and documented in the Mission Need Statement. The Milestone 0 review requesting approval to start developing potential design concepts occurs at the end of this phase. The most recent POM and/or BES along, with supporting rationale, are the only cost products required for the Milestone 0 review. Because the POM and BES are generally supported by rough–order–of–magnitude cost estimates, the CBRG review is aimed at validating the overall scope and magnitude of the resources required to implement the program.

After the program receives approval, the process of identifying and evaluating potential alternative ways of meeting the requirements takes place in the Concepts Development Phase. Costs and benefits of each potential alternative are evaluated and documented in the economic analysis (EA). Based on the results of the EA, the preferred alternative is selected and formal approval is requested at the Milestone I MAISRC review. In addition to the EA, the program manager's life cycle cost estimate and an independent life cycle cost estimate for the preferred alternative are required for the Milestone I review. The program's initial cost baseline is also established during this phase -- usually, but not always, it is some combination of the program manager's and the independent life cycle cost estimates.

The AIS is fully defined and designed during the Design Phase. The cost activities include updating the EA, the POM and/or BES, and the program manager's life cycle cost estimate. An independent life cycle cost estimate is also required at the Milestone II MAISRC review. In addition, any special program resource estimates (e.g., Selected Acquisition Reports) are also validated by the CBRG.

The approved system design is developed and fully tested during the Development Phase. The cost products for the Milestone III review are the same as those required for the Milestone II review -- they are updated to incorporate any changes identified during testing and other fact of life changes to the estimates. In addition, the CBRG also validates any trade-offs that may have been made to effectively balance cost, schedule, and performance.

The Deployment Phase encompasses the implementation of the new AIS at all of the operating sites. There are no Milestone reviews during the Deployment Phase.

In the Operations Phase the user operates and maintains the AIS. The chief costing activities consist of collecting actual operating and support (O&S) cost data and estimating the cost of any necessary system enhancements. Two MAISRC reviews (i.e., Milestones IV and V) are conducted during this phase.

- The purpose of the Milestone IV review, which occurs one year after deployment, is twofold. The first is to assess system operations to determine how well the AIS is meeting the goals established prior to deployment including affordability and benefit goals. The second objective is to approve any post-deployment modernization plans.
- The Milestone V review takes place approximately halfway through the operational life of the AIS, but no later than four years after Milestone IV. Its purpose is to determine if the AIS still satisfies the mission needs, requires modernization, or should be terminated.

The MAISRC cost products and when each is required are summarized in Fig. 4.

Schedule Considerations — Too often, the time required to properly prepare the MAISRC cost products is underestimated. The result is poorly prepared estimates that are unlikely to withstand the degree of scrutiny required for MAISRC reviews. To satisfactorily complete the MAISRC cost documentation and review requirements takes considerable time and effort by not only the cost estimator, but the program manager and other functional people as well. As illustrated in Fig. 5, the estimating process can take up to 145 working days, depending on the number of intermediate reviews required. This means that the estimating tasks need to begin approximately 7 calendar months (assuming 22 working days per month) prior to the MAISRC review.

Milestone	Economic Analysis (EA)	Program Office Estimate (POE)	POM/Budget Submissions	Independent Cost Estimate
Need Justification 0	No*	No*	Yes	No
Concepts Development I	Yes	Yes	Yes	Yes
Design II	Update	Update	Update	Yes
Development III	Update	Update	Update	Yes
Deployment	No	No	No	No
Operations IV & V	No	No	No	No

* All available cost and benefit data is required

Figure 4   MAISRC Cost Product Requirements

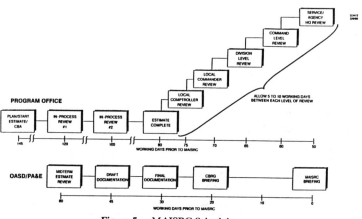

Figure 5   MAISRC Schedule

ESTIMATING DIFFERENCES – WEAPON SYSTEMS VS. AIS

This section discusses some of the ways that estimating automated information system cost differs from estimating weapon system. The differences included are drawn strictly from my own personal experience and are aimed at the experienced weapon system estimator trying to make the transition to estimating AIS costs.

Unit Price Behavior — Probably the first difference that a weapon system cost estimator notices is the behavior of hardware unit prices over time. The automatic data processing equipment industry is what is known as a decreasing cost industry. That is, over time the cost per unit of computing power is decreasing instead of increasing (see Fig. 6). The impacts of this phenomenon are illustrated in Fig. 7. Weapons system estimators are used to seeing constant dollar unit costs remain steady, or at best decline slightly over time, but the escalated dollars (the price actually paid) almost always increase (see Fig. 7a). In the AIS business, however, just the opposite is often true. It is common for escalated dollar unit costs to remain steady and, in

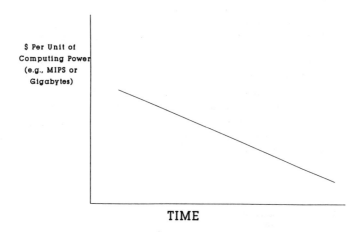

Figure 6   Computing Power Cost Trend

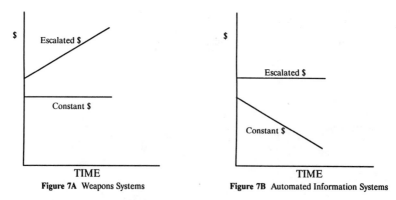

Figure 7A Weapons Systems   **Figure 7B** Automated Information Systems

Figure 7   Unit Price Behavior

many cases to actually decrease over time (see Fig. 7b), even though more computing power is being purchased.

Hardware Differences — Automated information systems generally make extensive use of off-the-shelf hardware which can usually be costed using vendor catalogs or existing government contracts. Weapon system hardware, on the other hand, is typically developed specifically for a particular application; therefore, catalog prices are nonexistent and contract prices are used only as a point of reference for an analogy estimating approach. Another consequence is that the AIS estimator is more interested in **quantity discount** effects rather than the traditional **learning curve** effects so familiar to the weapon system estimator.

Software Differences — In the area of software development there are significant differences in reliability and platform requirements. The consequences of a failure in weapon system software are generally much greater than for AIS software. For example, if a missile's guidance system fails to operate properly, the wrong target could be destroyed and the results could be disastrous. The results of AIS software failures are generally measured in terms of late reports — seldom, if ever, are human lives at stake. Therefore, software reliability requirements are much higher for weapon systems than for AISs.

Operating and Support Phase Differences — The estimator needs to be aware of several differences in estimating O&S costs. First, the life cycle is usually 8 years (sometimes less) for AISs while 15 or 20 years (even longer in some cases) is common for weapon systems. Second, the maintenance concept for weapon systems is usually organic while AISs are usually maintained by contractors. Finally, the O&S cost drivers are different. Weapon system O&S costs are usually driven more by maintenance considerations such as failure rates and repair times while AIS costs are driven more by the amount of electricity used and the number of system operators required.

REFERENCES

1.   AFCSTCP–1, "Major Automated Information System Cost Estimating Reference and Guide," March 1989, The Air Force Cost Center.

2.   DoD Directive 7920.1, "Life Cycle Management of Automated Information Systems," 20 June 1988.

3.   DoD Instruction 7920.2 "Automated Information System (AIS) Life–Cycle Management Review and Milestone Approval Procedures," 7 March 1990.

4.   "DoD Cost/Benefits Guide," 30 May 1989, OASD (PA&E).

The Complete COCOMO Model: Basic, Intermediate, Detailed, and Incremental Versions for the Original, Enhanced, Ada, and Ada Process Models of COCOMO[1]

Ronnie E. Cooper

HQ AFLC/FMCR, Wright-Patterson Air Force Base, OH 45433

ABSTRACT

Since its publication in 1981, the COCOMO model presented in Software Engineering Economics (SEE) by Barry W. Boehm has been at the forefront of software models. Since 1984, the existence of the COnstructive COst MOdel (COCOMO) User's Group (CUG) has served to maintain the needed information exchange and to be the vehicle for subsequent updates to the COCOMO model (by Dr. Boehm). Its public acceptance is a matter of record. The Software Engineering Institute (SEI) has served as the sponsor of the CUG since about 1987. The Department of Defense and Cost Organizations both have strong interest in using the best methodologies available for software costing. COCOMO is the best documented such method and has a wide range of uses. The COCOMO model has promoted the purposes of Software Engineering since before 1981. It has not become dated. It has more than 20 automated implementations. In fact, many people are still discovering this model.

Three enhancements have been released by Dr. Boehm. Each of these are discussed to summarize their effects and uses. This paper is a limited presentation of the full details of Dr. Boehm's CUG presentations and contain key portions from a forecoming book on COCOMOID by this author. COCOMOID is a LOTUS 1-2-3 version available free of any license fee. It is a complete implementation of all aspects of COCOMO that have been released into the public domain, plus many useful analysis enhancements to speed use and understanding of the COCOMO model while being used.

---

[1] Featuring COCOMOID, version 3.2, a LOTUS™ 1-2-3 worksheet unlike any other for commercial and defense industries use.

The estimators reason for being - good reliable reasoning.

"For which one of you, when he wants to build a tower, does not first sit down and calculate the cost, to see if he has enough to complete it? Otherwise, when he has laid a foundation, and is not able to finish, all who observe it begin to ridicule him, saying 'This man began to build and was not able to finish.' Or what king, when he sets out to meet another king in battle, will not first sit down and take counsel whether he is strong enough with ten thousand men to encounter the one coming against him with twenty thousand? Or else, while the other is still far away, he sends a delegation and asks terms of peace."
Luke 14:28-30 [Bible, New American Standard Version]

It is the purpose of this paper to help the reader discover the **complete** COCOMO model[2]. That is to say, beyond the book, <u>Software Engineering Economics</u> [SEE], by Dr. Barry Boehm, copyright 1981. The COCOMO Model has been updated and expanded by three enhancements. And not everyone has read all 729 pages of [SEE]. If you have read all 767 pages (including the index), you understand. This paper will simplify explanations and concentrate on the basics of COCOMO. All tables, definitions, and processes in [SEE] will not be repeated here. Elaboration will be confined to the new options, tables, and definitions. First, let's outline the extent of the COCOMO model, then briefly cover some ways to use the model, and finally discuss a sample implementation of COCOMO - COCOMOID[3].

I. The Extent of the COCOMO Model

The COCOMO model is a **set** of mathematical models[4] in the form ax^b . Here 'X' is defined as Delivered Source Instructions[5] in Thousands (KDSI), 'a' is a linear

---

[2] Since 1984, the COCOMO User's Group has served to maintain an educational process and to distribute improvements to the model. The Software Engineering Institute (SEI) has served as the sponsor since 1987.

[3] COCOMOID.WK1 is the most complete spreadsheet version this author is familiar with. But a multitude of people have used the spreadsheet to create their own quickly developed versions of COCOMO to check work or develop understanding. Over 20 other commercial or DoD versions exist, each with their own strengths and uses. Portions of this paper are from a fore-coming book on COCOMOID by the author.

[4] See appendix A for the nominal effort equations of the original COCOMO. Other models of COCOMO are Enhanced, Ada, and Ada process.

[5] Also referred to as Lines Of Code (LOC) in a more loosely defined context. Delivered is defined to exclude nondelivered support software such as test drivers, unless they are developed with their own reviews, test plans, documentation, etc. Source instructions include all program instructions created by project personnel and processed into machine code by some combination of preprocessors, compilers, and assemblers. It excludes comments cards and unmodified utility software. It includes job control language, format statements, and data declarations. Instructions are defined as card images (SEE, pgs 58-59). COBOL and Ada have differences from these definitions. COBOL counting excludes 2/3 of the nonexecutable source statements (SEE, pg 479). Ada counts the number of carriage returns in package specs and the number of nonembedded semicolons in package

multiplicative composite of several factors[6], and 'b' is an appropriately chosen exponent between 1.04 and 1.24[7]. The several factors affecting 'a' are the 19 attributes listed in appendix B and the calibration value.

Each COCOMO model has four versions: basic, intermediate, detailed, and incremental development. Each of these are increasingly accurate using the original 63 projects as the standard of comparison. The incremental development is most accurate only if the project was really developed using incremental development design.

Each version of COCOMO has three modes; organic, semidetached, and embedded. These modes represent the degree of difficulty a software project expects to have, particularly in communication problems and project size. See appendix A for more detailed information on modes.

The COCOMO model estimate limits itself to direct evaluation of the four phases of the waterfall model[8] shown in boldface below: feasibility, plans and requirements, **product design, detailed design, code and unit test (programming), integration and product verification (test)**, implementation and system test, operation and maintenance, and phaseout. It also limits itself to eight activities: requirements analysis, product design, programming, test planning, verification and validation (V&V), project office functions, configuration management and quality assurance, and manuals. All other activities must be factored in.

The COCOMO model was designed to assist and quantify the software engineering processes. It models a real world set of these processes and quantifies them into a

---

bodies, while excluding comments. The reuse model is used with all (Ada COCOMO: TRW IOC Version, 4 Nov 87).

[6] Simply put, multiple a bunch of positive numbers together to get the total effect on nominal effort.

[7] If these values seem different from the book (SEE), they are. They expand the model and are used with one of the extension, the Ada process model. See appendix C for more detail on these values.

[8] Chapter 4, SEE

reasonable estimate of man-months of effort and schedule. It is important to realize the range of equations represented by COCOMO. They cover the processes from the earliest rough order of magnitude to the later more defined stages. This allows estimates to be further refined as more information becomes available later.

As such, the appropriate form of COCOMO must be chosen for the project scope, timing, and the method of development being used. The simplest form is the Basic version using only KDSI. The Intermediate and Detailed versions add attributes and phase[9] dependent relationships respectively, while using different coefficients and exponents within their equations. The Incremental Development version considers a large project to be broken into smaller pieces with each piece separate, but building on prior pieces. When all the pieces are complete, the project is complete. Each piece is a development effort under one master plan. The Incremental Development version uses the Intermediate attribute tables.

The Intermediate model is the most widely used version of the four. This is because it obtains good results and it is simple to do manually or with a spreadsheet. Ratings are assessed for each of the 18 development COCOMO attributes and the respective table values are then looked up for each rating. These values add or reduce effort by using varying values near "1.00" which are multiplied together to get the total effort adjustment factor (EAF) effect (e.g., .93 x 1.04 = 96.73 EAF, if all others attributes are one). After a total adjusted man-month effort is calculated, phase tables are used to separate the effort into phase pieces. The schedule is a function of the effort and, if off-nominal values are used for SCED, the schedule attribute.

All related parts of a project are evaluated together as a whole. Multiple sub-projects that are part of the total project are used together. The project should not be broken into separate pieces for separate evaluation unless they are truly independent.

---

[9] The four phases are: 1) plans and requirements, 2) product design, 3) programming, and 4) integration and test, [SEE], chapter 4.

There is a maintenance version like the intermediate version. It eliminates the effect of one of the 18 development attributes (SCED), and adds one additional for requirements volatility (RVOL) adjustment (normalization)[10]. Two more of the 18 attributes (RELY and MODP) are logical inverse values of their development values (wise development investment costs reduce maintenance costs, except for very high RELY). When maintaining developed code, the LOC should be counted differently than during development, potentially including reuse code. The total KDSI of code that must be maintained is the logical choice, since all design, code, and integration needed must be reviewed and understood to be maintained. Some reusable code that was developed for reuse at a very generic level may have no maintenance other than learning what it is supposed to do and ensuring it is properly used, hence its code may not be counted, or counted at reduced effort. The COCOMO calculation process for maintenance is the same as development, but with different values being used. Evaluate the environment maintenance is to be done in, and consider the analysis of other factors to select the proper input values. The logical differences between development and maintenance of how, when, and how long maintenance will be done may require different attribute choices than when the project was developed. Also, the initial heavier maintenance work load just after software is first released does need to be considered and weighted into the maintenance workload estimate.

The development COCOMO attributes are 18 of the 19 defined characteristics with different assigned numerical values. The nineteenth will be discussed separately later. See appendix B for a sample set[11] of these values and graphic comparisons of the four different sets for each model of COCOMO. Attributes are used to show the effect of various other factors beyond size (KDSI) and mode (inherent difficulty). Each attribute has different values assigned to it, therefore, they have varying effects on total effort. The ratings for each attribute varys from Very Low to Extra Extra

---

[10] SEE, pg 550

[11] Two attributes, ACAP and PCAP, are interpolated in the Ada process model and hence vary within a range.

High[12] depending on the model. An important difference with the detailed model compared to the intermediate model for attributes is that there is a set of values for each of the four phases of development. Hence, the detailed model varies effort-by-phase, and the resulting total effort for the project will be very different from that using nominal attribute inputs and can be considerably different from results using the intermediate model with the same non-nominal ratings. The use of non-nominal inputs implies management "gray matter[13]", resources, and time are being used effectively for good inputs, or are being used with some limitations with the more restrictive (costly) inputs.

The detailed model uses the same process as the intermediate model for each of the four phases covered directly by COCOMO in a hierarchical of three things; the entire system, the subsystems, and the modules. Each hierarchy uses the phase sensitive effort multipliers and four phases (this gives four effort calculation to be summed for a total) rather then one long phase as in the intermediate version (which is broken into phases after determining total effort). This process makes the detailed model about eight times more involved than the intermediate, but slightly more accurate.

There is a calibration model which, like the maintenance model, also uses the 19th attribute to account for requirements volatility (RELY). The total effort of a project is reduced by using the requirements volatility input as a divisor (it's value is greater than one since it reduces) to the project effort. This result achieves the nominal normalized effort the calibration is then done on. All COCOMO calibration effort **assumes good management**, etc. to obtain **nominal results**

---

[12] Very Low, Low, Nominal, High, Very High, Extra High, Extra Extra High are the complete rating set. Each is defined in detail in [SEE], e.g. pages 118-119, or chapters 24-27.

[13] Management time to think about and periodically review the process. The more factual information and trained personnel are used to make decisions, the better overall project results can be. Marketing and bidding information may have a place in the decisions, but results will depend on the hard facts of the business capabilities on hand and in use. Other management decisions will have to pay their own way, because they also have their own costs for the "benefits" management may desire from them. It is very dangerous to the bottom line to make COCOMO environment inputs (global, size, and attributes) that reduce costs that do not have management understanding and committment to the costs and efforts needed for such inputs to be real.

Therefore, excessive requirements volatility is always eliminated from actuals for calibration (and maintenance). RVOL use removes the effect of code that was throw away during development, therefore the true nominal effort is what code was needed to do the requirement (the ratio "work done/work required" is always greater than or equal to one). Nominal workload is defined as a multiplier of one. Exceptional management is covered in attributes and methods of development for purposes of predictions. Therefore, all calibration values are assumed to be for nominal normalized efforts, .i.e., after all effects of all attributes are considered.

Some implementations have made the requirements volatility attribute part of the development models, but Dr. Boehm says "this is a highly subjective, imprecisely defined parameter whose value would not be known until the completion of the project"[14]. In paraphrase, if we don't plan the changes, how can we know what they will be? Therefore, the logical conclusion should be to model just what is known, therefore leaving out requirements volatility (why plan for mistakes, if you can plan for then, you should be able to plan around them and eliminate them). Some amount of requirements volatility is built into the model values anyway. All projects are subject to it, IBM has estimated as much as 25% is common[15].

The original COCOMO model of 1981 has had three refinements.

THE FIRST REFINEMENTS

The first refinement to the COCOMO model was a paper by Barry Boehm and Walker Royce, TRW, Ada™ COCOMO: TRW IOC Version, proceedings, Third COCOMO User's Group Meeting, CMU Software Engineering Institute, Pittsburgh,

---

[14] SEE, page 484

[15] SEE, pg 484

PA, November 4, 1987. This paper introduced three categories of differences. It also expanded the brief prior description on the incremental development model[16].

I call the general improvements to COCOMO - the enhanced version. These comprise a wider range of ratings and effects due to software tools (TOOL) and turnaround time (TURN); the splitting of virtual machine volatility effect (VIRT) into host machine (VMVH) and target machine (VMVT); the elimination of added costs due to schedule stretchout (SCED), which is really the effect of of incremental development practices stretching schedule, but reducing effort (the assumption is again - use of good practices; the addition of cost drivers to cover effects of security-classified projects (SECU) and development of software reusability (RUSE); and the addition of a model to estimate the costs and schedules involved in using a phased incremental development process.

I call the effort and schedule effects specific to Ada - the Ada version. They include the enhanced version effects. Additional effects are reduced multiplier penalties for higher levels of required reliability (RELY) and complexity (CPLX); a wider range of ratings and effects due to programming language experience (LEXP); a set of Ada-oriented instruction counting rules, including the effects of software reuse in Ada.

The final model I call the Ada Process Model because of the effects to COCOMO of using the Ada Process Model and related practices. These can **also largely be adapted to projects using other programming languages**. These effects include the revised exponential scaling equations for nominal development effort, development schedule, and nominal maintenance effort; the extended range of Modern Programming Practices effects (MODP); the revised ranges of Analyst Capability (ACAP) and Programmer Capability (PCAP) effects; and the revised phase distributions of effort and schedule. This version has very wide robustness with 60

---

[16] SEE, pg 504

[17] Their use on non-Ada projects would require some experimental tailoring of standard COCOMO to accommodate the resulting effort and schedule effects.

different exponents possible plus the different attribute tables accumulated for all the enhancements above.

THE SECOND REFINEMENTS

The second refinement to the COCOMO model was the presentation paper by Barry Boehm, TRW, Ada™ COCOMO REFINEMENTS, proceedings, Fourth COCOMO User's Group Meeting, CMU Software Engineering Institute, Pittsburgh, PA, October 4, 1988. This paper introduced additional definition of the Ada process model and its management implications.

The Ada process model effects are four fold:

1) Ada process model milestones are detailed in outlines for each milestone. These milestones are SSR (Software Requirements Review), PDR (Preliminary Design Review), CDR (Critical Design Review), UTC (Unit Test Completion), and SAR (Software Acceptance Review).

2) Established Σ rating scale aids for determining Σ ratings at early stages. These show typical characteristics of rating scale levels for: Design thoroughness by PDR, Risk elimination by PDR, and Requirements volatility. Maintenance rating scale levels are for: Modern Programming Practices (MPP) used in maintenance, and Conformance to the Ada process model during maintenance. See Appendix C for these scale aids.

3) Noted effect of Σ on ACAP, PCAP requires interpolation for partial compliance with the Ada process model.

4) Lastly, warned of effect of Σ on phase distribution also requires interpolation for partial compliance with the Ada process model.

Interpolation for the cost factors (CF) [3, 4 above] uses a formula for ACAP, PCAP, Effort %, and Schedule % interpolation. See appendix D for these tables.

The management implications of the $\sum$ rating scales are summed up as lists of do's and don'ts.

The **do** list is:
1) Tailor the Ada process model to your organization.
2) Use a sequence of risk management plans to drive the software process.
3) Use incremental development to stabilize the development process.
4) Work to eliminate sources of uncertainty in requirements.
5) Use a small, top team to develop the architecture and resolve risks.
6) Build up experience with the Ada process model.
7) Provide strong tool support for the early phases.

The **don't** list is:
1) Force premature requirements and design completion milestones.
2) Swamp the architecture definition process with paperwork or with large numbers of mediocre people.
3) Use document milestones to drive the project's organization - e.g., establishing a separate "requirements team" and a "design team".
4) Force every development into the same sequence of steps.
5) Try to build the whole system all at once.
6) Wait around passively for someone to provide you the definitive requirements.

THE THIRD REFINEMENTS

The third refinement to the COCOMO model was the presentation and paper by Barry Boehm, UCLA, <u>Recent Developments in Ada</u>™ <u>COCOMO</u>, proceedings, Fifth COCOMO User's Group Meeting, CMU Software Engineering Institute, Pittsburgh,

PA, October 18, 1989. This paper introduced the organic and intermediate modes to Ada with the prior embedded mode. Discussion of the Ada process model strategy and more detailed comparisons with the original COCOMO are also presented.

The Ada COCOMO Development-mode models are:

MODE	EFFORT	SCHEDULE
ORGANIC	$MM = 1.9 \cdot \prod(EM) \cdot (KDSI)^{1.01+.25\Sigma}$	$T_{DEV} = 3.0 \cdot \prod(MM)^{.38+.2\Sigma}$
SEMIDETACHED	$MM = 2.35 \cdot \prod(EM) \cdot (KDSI)^{1.02+.625\Sigma}$	$T_{DEV} = 3.0 \cdot \prod(MM)^{.35+.2\Sigma}$
EMBEDDED	$MM = 2.8 \cdot \prod(EM) \cdot (KDSI)^{1.04+\Sigma}$	$T_{DEV} = 3.0 \cdot \prod(MM)^{.32+.2\Sigma}$

These are based on an Ada database of: 12 embedded projects and 6 organic projects. Twelve of these were non-TRW[18] projects, 6 were TRW projects.

For the equations, accuracy measures are:

Modes	Effort		Schedule	
	R	P(20)	R	P(20)
Embedded	.207	.67	.163	.71
Organic	.191	.50	.395	.60

The net result of these accuracy measures is continued data collection and analysis is still called for. Continued dual use of the original COCOMO is still advised based on these results.

---

[18] Dr Boehm was chief scientist at TRW for several years and would have excellent access to TRW information.

II. Some ways to use the COCOMO model

The obvious use of COCOMO is to estimate the effort and schedule of a software project COCOMO is designed to evaluate[19]. From these estimates, personnel manning and project costs are short additional steps. Given the riskiness involved in software estimates, the adjustments available for size and model risk are prominent considerations all project management must measure and model for possible inclusion in their decision process.

Other obvious reasons to use the original COCOMO model in different ways are the new refinements in COCOMO, e.g., Ada and the Ada process model in conjunction with the incremental model. There are many ways software is developed, just as there are many ways hardware is developed. It is only reasonable to model the software the way it is designed. Hence, the COCOMO models are sets of options to try to best model the many development environments.

The "What-if" use of a model is another good use to help us understand a model and evaluate alternatives for management decisions. Comparative views, particularly of the intermediate and detailed COCOMO versions, demonstrate the real benefits some decisions can lead to in reducing cost, etc., and to show the difference between the models in practical use.

Within any organization additional reasons to use COCOMO are;

1)   To develop internal expertise for the organization using the openness of the COCOMO model to understand each step;

---

[19] Given that COCOMO has a wide potential range of mathematical models, calibrations, wide spread experience or training in companies and schools, published specifications, numerus implementations of the model on the market, and widespread acceptance of its use - it is applicable or adaptable to almost any software situation.

2) To make comparisons with other models data inputs as cross-checks and to lead analysts to discover the differences and similarities of the models;

3) To allow model customization/optimization.

Continual use of COCOMO leads to further value. Estimates can be iterated as more information becomes available to update estimates. Data sets and documentation can be saved and evaluated for the effects of real changes or projected management decisions.

After project completion several considerations should be planned. With milestone or final data availability, specific calibrations can be obtained. If careful data collection is done, attribute tables can also be customized. Final reports should comment on the accuracy of requirements and estimates as the project evolved. Careful use should build management confidence. Where errors are apparent, corrective actions should be taken and documented. Knowing (measuring) the cost of those errors (hence the value of prevention) should be a very high priority with management and data collection/analysis given that same priority. A long term strategy of managing and valuing information prevails over the quick and dirty strategy of ignoring the capabilities, skills, and resources of a company represented in data not collected. Care must be taken to measure results and take directed actions based on the best **information** available. Experience is expensive, so don't throw it away or ignore it by not collecting or managing it.

Calibration data[20] should be easy to use for any COCOMO version used. Its availability from other sources should allow duplication of results from those sources. Most importantly, calibration must be used in any long term software management strategy.

---

[20] An excellent tool by SoftStar, "Calico", has been distributed without license fee. It allows extensive evaluation of COCOMO calibrations and also development of new calibrations using your own data with or without the provided COCOMO data.

III. The sample implementation, COCOMOID.WK1

In discussing this sample implementation, we will:

1)   cover the extent of the COCOMO model COCOMOID accomplishes,

2)   point out user interface features that ease use,

3)   discuss support files used,

4)   discuss building a data file,

5)   demonstrate how to use key features, and

6)   show some sample outputs (of the many available).

What extent does COCOMOID accomplish the COCOMO model

The COCOMOID model handles the full extent of the COCOMO model. It operates in all modes on all versions, many of them simultaneously. It supports integrated development and maintenance life cycle cost (LCC) estimates in one data effort. All 19 attributes plus an extra one for future use are completely ready to use. All equation values are available as defined by the model. Inputs are integrated into logical groups to allow for all versions: basic, intermediate, detailed, and incremental development. The calibration model for the coefficient values are calculated using actual nominal effort inputs. The resulting values are savable in up to 50 separate calibration table files, each with up to 42 values (14 values in each of the three modes, a result of 2100 calibration values possible). It is limited to handling one project (CPC) of up to 14 subsystems (CPCIs) or modules or units at one time. But multiple data sets can be handled and these can build up very large data sets. The Original, Enhanced, Ada and Ada process model can be used separately and with the Incremental Development version.

COCOMOID user interface features that ease use

A comprehensive menu driven user interface allows all features to be available to ease use of the model. The spreadsheet formulas are not changed by the user. COCOMOID operates as an engine on the data. Its menu system is, in fact, password protected so accidental changes don't happen, but controlled changes are easily possible. It is available as a compiled version so LOTUS 1-2-3 is not required for use. It is modular. Another worksheet can be used to build input sets, if additional quantitative assistance with inputs is desired.

COCOMOID is a fully integrated model for all versions of COCOMO. Inputs are required only once to handle all versions. Global inputs are consolidated. Development and maintenance inputs are both used so a complete life cycle cost is possible. All tables and interpolations are done as needed. Over 400 error checks are done on inputs to assist data input, with error messages displayed on screen throughout to identify the nature of the logical input error.

The COCOMOID support files used

Since COCOMOID is a self contained engine, it uses several external support files to maintain flexibility for the different uses as grouped below.

1)   The most important supplied files are the **detailed attribute tables** for each version. Each table is a password protected file to prevent unwanted changes.

2)   **Calibration files** containing different tables[21], each with up to 42 values and notes, support most any calibration need. With 50 files, that's 2100 different

---

[21] Each is identified as CAL_.WK1 with the blank being the unique identifier. Identifier can be A thru Z, 0 thru 9, and the 14 special characters: ! @ # % ^ & _ - ~ ' ' " { }. This is the 50 tables allowed.

values. These are also password protected with full control of contents from within COCOMOID.

3) All **data sets** and their related documentation are separate files with the same name and a "DAT" or "DOC" extension. The data files are handled easily and quickly to allow fast multiple data runs and comparisons. Data files also have their names embedded within the file to maintain integrity, but are not password protected.

4) For ease of use a small, standard **printer configuration** file is used to set up COCOMOID each time it is run (as it was last saved).

5) All 18 **graphics** can be saved for each data set using the last four characters of the dataname and four unique COCOMOID characters for each of the 18 graphs.

6) The last set of files are the **user help files**. The help files are not protected and are standard worksheets. This allows users to customize the help file to suit their needs. One file is reserved totally for the purpose of unique user help information.

A COCOMOID sample data file

Text 1, is a sample input form for the Global inputs in COCOMOID, Version 3.20. The underlined entries are the inputs. Appendix E contains blank COCOMOID input forms. Each of four sections (Globals, Titles, Sizing, and Attributes) will be discussed separately.

The first section, global inputs, has eight major areas: 1) Identification, 2) Man-month default, 3) Calibration choice, 4) Options, 5) Rates, 6) SCOW (Skill Coding Of Wages), 7) PERT (Program Evaluation Review Technique), and 8) Maintenance.

The Identification area, Text 2, is needed for the name of the data set, the COCOMO version and mode to use, and for what the Ada process model inputs are, if applicable. The blocks headed "Exper/MPP" and "Dsn/Cnf" are for double entries, the left number is for development and the right number of the double numbers is for maintenance. "Exper/MPP" is **Exper**ience with the Ada process model

```
COCOMOID 3.200      Extra <RETURN> stops Global input   DATE-> 27-Jun-91
  Data only name>   FEWSUNIQ <Proj 2,What-ifs 2             0 LOGIC ERR's
                                                          168 <Mnhr/Mo152
COCOMO Version >          2  0,Bk;1,Enh;2,Ada;3,Proc       19 <Mndys/Mo19
Development Mode>         3  1=Org,2=Semi,3=Embedded
SEE Ada detail:    Exper/MPP Dsn/Cnf Risks. Req Vol.
Ada process, 0-5>        44       44        4        4  $0 Feasibility
                                                       Alt Mo $ Activity
  Calibration 0  >       14  0=Std,1-14=input table|   $7,020 Req Anal

LOC Opt In      >         0  0=In,1=Min,2=Exp,3=Max|   $7,020 Prod Dsg
Model Risk Incl>         0  0-3, # * 20%                $7,020 Progm
SCOW $ Use = 1  >         0       0 <Add PERT Risk      $7,020 Test Pln

Programmer SCOW    Starting  Median  Top   Base Yr     $7,020 V&V
  Salary range>   $27,000 $42,800 $71,500    1990      $7,020 Proj Off
Salary SCOW             O/H     G&A  Profit Infl BY     $7,020 CM/QA
Loading Factors>    130.0%   30.0%  10.0%    27.6%      $7,020 Manuals

PERT Min %      >       25% 0%-50% Range, 25% Rmd       10.0%Overtime%
                       a Rmd 3,m Rmd 2,b SDs,0-3
PERT Weights    >        1       3       2        0|   % Mnt Profile 3 yr
                                                       100.0% 1st yr
                   a>8# Yrs $ Diff% IMEM.4-2.5,Pg540   100.0% 2nd yr
Maintenance     >       16       0%  1.09             100.0% 3rd yr
```

Text 1 - Sample Global Inputs with input area separation emphasized

in development and Modern Programming Practices (**MPP**) used in maintenance. "Dsn/Cnf" is for **D**esign thoroughness at development Preliminary Design Review (PDR) and

```
COCOMOID 3.200      Extra <RETURN> stops Global input
  Data only name>   FEWSUNIQ <Proj 2,ID 4,What-ifs 2

COCOMO Version >          2  0,Bk;1,Enh;2,Ada;3,Proc
Development Mode>         3  1=Org,2=Semi,3=Embedded
SEE Ada detail:    Exper/MPP Dsn/Cnf Risks. Req Vol.
Ada process, 0-5>        44       44        4        4
```

Text 2 - Identification area

Conformance to the Ada process model during maintenance.

The default man-month area, Text 3, needs entries in either hours or days, but man-hour entries > 152 take priority. Setting values for man-days requires an entry > = 19 and man-hours < =152. The normal values are shown on the right and are the values COCOMO assumes for the standard man-month. A value

```
DATE-> 27-Jun-91
  0 LOGIC ERR's
168 <Mnhr/Mo152
 19 <Mndys/Mo19
  ^ Chg only 1
```

Text 3 - Man-month Defaults

larger than 152, e.g., 168 means either a company is expecting 168 hours of work each month for standard salary or that it is expected that 16 hours or less of paid overtime will occur each month. If the time over 152 hours or part of that time is overtime then the percentage should be entered in the overtime cell. If in this case 8 hours of the "standard man-month for scheduling" 168 hours is paid overtime, then 8/168 or

4.8% is overtime. Companies sometimes use unpaid overtime to "get well" because of "buying in" or under estimating on a contract. Excessive use of unpaid overtime is a tipoff of potential project problems. Excessive hours are hours well over the standard 152 hours by **any team** on the project. Sustained manpower use at these high levels will also lead to potential burnout and excessive bugs in software even if they are paid overtime. Carefully consider this variable and its consequences to later analysis.

```
Calibration 0  >        14  0=Std,1-14=input table|
```
Text 4 - Calibration table choice

The calibration option area, Text 4, is entered as a number between one and fourteen for the table in use which is noted right after the word "Calibration" as a single character ("0" in the case, the default table). All calibration tables contain all three modes, so the development mode input identifies which is used within the table. Each table can contain 14 sets of values. A set is one value for each of the three modes and descriptions for each.

The Options area, Text 5, contains four important options explained below. They can be defaulted to have no extra effect by entering a zero in each. Each of

```
LOC Opt In     >    0   0=In,1=Min,2=Exp,3=Max|
Model Risk Incl>    0   0-3, # * 20%
SCOW $ Use = 1  >    0        0 <Add PERT Risk
```
Text 5 - Options area

these option inputs are toggles, inputting "0" turns off the option and inputting "1" turns the option on.

The "LOC Opt In" input allows four fast "what-ifs" calculations to range the estimate; input, minimum, expected, and maximum values for LOC. After all inputs are made, changing just this one value allows complete recalculation at these value points. The values used can be viewed or output and the original inputs are not destroyed.

The "Model Risk Incl" input option specifies how many standard deviations to add to results to account for the known COCOMO model risk. In COCOMO, one standard deviation adds 20% to the estimate.

The SCOW option, "SCOW $ Use = 1", when turned on by inputting "1", does what the acronym stands for, it is used for **S**kill **C**oding **O**f **W**ages (SCOW). SCOW is a unique COCOMOID method using the personnel attribute inputs for skill and experience to estimate a corresponding appropriate wage. This is an excellent tool to check reasonableness of input monthly programmer cost values.

The final option is to "Add PERT [size] Risk" to the final answers using "1" to turnthe option on, or leave answers at nominal values using "0". The amount added depends on the inputs in the PERT area. If "SDs,0-3" in that area is "0", nothing will ever be added for PERT size variance risk (the input is multiplied by 20%, therefore, 0 x 20% = 0).

The Rates area is for optional dollar related inputs. The first is a pass through feasibility cost input, Text 6. The other two parts

$0 Feasibility

Text **6** - Feasibility Study Cost Input

are for another optional method of costing using fully burdened monthly costs by type of activity, and last, an input for the percentage of effort that is done using overtime.

Next are optional inputs for **fully loaded**monthly costs, Text 7, for each listed type of activity. This alternative allows costing to be done where these are the rates known.

Alt Mo $ Activity
$7,020 Req Anal
$7,020 Prod Dsg
$7,020 Progm
$7,020 Test Pln
$7,020 V&V
$7,020 Proj Off
$7,020 CM/QA
$7,020 Manuals

Text **7** - Activity rates

10.0%Overtime%

Text **8** - Overtime %

Finally this area allows a percent of effort done on overtime, Text 8. This increases total costs for the stated percentage of effort at a 50% higher rate. An error message will occur if the percentage of overtime is greater than then the standard man-hours available for overtime (using 152 as the base) and the maximum percentage allowed will be shown. The maximum percentage is $\left[\dfrac{(\frac{Mnhrs}{month}-152)}{152}\right]-1.0$, e.g., 160 hrs/mo

is [(160-152)/152]-1.0 = 5.26%.

The **SCOW** area, Text 9, is unique to COCOMOID. This optional area is used for costing monthly costs only when turned on in the options area. Salary

Programmer SCOW	Starting	Median	Top	Base Yr
Salary range>	$27,000	$42,800	$71,500	1990
Salary SCOW	O/H	G&A	Profit	Infl BY
Loading Factors>	130.0%	30.0%	10.0%	27.6%

Text 9 - SCOW Inputs

range is used to input the average each grouping would expect to receive as a salary. The other inputs convert these to fully loaded (burdened) rates. An algorithm converts these inputs into 21 pay rates to be selected according to personnel attribute inputs.

The PERT area, Text 10, also has some unique things in COCOMOID. This area has a simple percentage input for minimum code size that is applied against

PERT Min %	>	25% 0%-50% Range, 25% Rmd
		a Rmd 3,m Rmd 2,b SDs,0-3
PERT Weights	>	1 3 2 0

Text 10 - PERT Inputs

actual input ADSI and maximum ADSI to determine logical minimum LOCs. This algorithm below for minimum ADSI balances the effects of how close input ADSI is to zero against the riskiness based on the range between input and maximum ADSI. It is the maximum of either:

Minimum ADSI = Input ADSI - PERT Min % * (Maximum ADSI - Input ADSI)

or   Minimum ADSI = Input ADSI * (1 - PERT Min %)

The PERT weights are the suggested distribution for minimum (a), most likely (m), and maximum (b). This suggested weighting is for the expected probability of costs being greater then expected rather than them being less than expected. This left skewed distribution is more realistic than a balanced one like the normally used 1-4-1 one. Input the number of standard deviations to add, based on size results, using the "SDs,0-3" cell in this portion of area. The bias COCOMOID uses here is intended to give a measure to size under-estimation and the dual use of a calculated minimum ADSI and of an asymmetrical distribution do this. They can be evaluated to suit risk evaluation or turned off completely, but the results in either case will be shown as a percentage that is added to results. In COCOMOID, to remove mean/mode distribution bias, set "a" = "b" = 1.

The final area is for maintenance. Four types of entries are present. The number of years for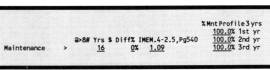

Text 11 - Maintenance parameters

maintenance is a straight multiplier for the calculated annual maintenance. The dollar difference percentage is based on development rates. If lower labor rates are expected, a negative percentage for that difference is entered. When higher maintenance labor cost rates then those during development are expected, then a positive percentage of increase is input. All dollars are constant year dollars. The IMEM (Ideal Maintenance Effort Multiplier) has a range in [SEE] from .4 to 2.5 with 1.09 its computed value.

The last three percentage inputs allow a phase in of maintenance costs. All values should be greater than or equal to one. Logically they should not increase from top to bottom (that indicates development, not maintenance).

The second section of inputs covers identification and titles of the CPCI/CPC's in Text 12. Three identifications are used: a name, a numerical increment number for incremental development of that name, and a numerical subsystem number for the CPCI/CPC to indicate which CPCI it belongs in within the project. In a simple example, all numerical entries will be ones, "1". If increments numbers are left out, the "IC" column is assumed to be "1's". Subsystem numbers must be input as a value from "1" to "14", with duplicates allowed in any order. A subsystem number of "0" or blank indicates there will not be any further data entered on that row. Any entry on a row will be considered an error until the row is given an "IC" number. This is forced because subsystems are treated differently than modules in the detailed model.

The numerical indicator used for each row is a integer for identifying which "IC" increment ("1" through "6") and which "M" subsystem ("1" through "14") a row of CPCI/CPC inputs belong to in the project. A number is input next to the identification titles for each row. Text 12 is an example where there is only one increment and one subsystem for all data entries, .i.e, only one integer, "1", is used for numerical identification.

Rg0-14 for M# --v		
Component	IC	M
Mis E/F Proc1		1
Oper Sup	1	1
Sim & Test	1	1
Training	1	1
Sup Tools	1	1
Dev&Mnt Sup(1		1
Sys Sim Sup	1	1
Sys Test	1	1
PL1 Tst&Intg1		1
Factory Sup	1	1

Text 12 - CPCI/CPC Titles & Identification

Most of the COCOMOID inputs are straight out of [SEE] for project specific inputs with three exceptions using numerical indicators for identification: 1) The form of attribute inputs, and 2) the increment identification and 3) the subsystem identifier for each row of CPCI/CPC inputs belonging to a project. The last two of these exceptions were just discussed in this section.

Mod Spread -->		Adapted	40%	30%	30% Cnvs Pln	
FEWSUNIQ		Del Src	Design	Code	Intg Inc,Pg558 Max>	ADSI
Rg0-14 for M# --v		Instruction	(0-100)	(0-200)	(0-6)	(#>ADSI)
Component	IC M	ADSI	DM	CM	IM CPI	MADSI
Mis E/F Proc1	1	72200	100	100	100 0	108300
...						

Text 13 - Development sizing area

The third section of inputs covers sizing and a monthly programmer cost input. Text 13 is a set of examples of a single CPCI entry with inputs underlined of the fourteen rows allowed.. This first part of two parts is for code value sizing entries and is actually the left half, the second portion is immediately to its right. It is separated for clarity and consistency with COCOMOID screens. The Adapted Delivered Source Instructions (ADSI) is the most likely size ("m" value in

PERT) in LOC form. These inputs gage the size of the parts of a project or the whole project in LOC and related factors. The section columns DM, CM, and IM[22] are for percentage inputs by category. Inputs of "100" means 100% of this category is passed through. Integration inputs can be up to 200% because they involve more then just the code being input (there is system integration that needs to added). The Conversion Planning Increment (CPI)[23] is a percentage additive used when code is converted between languages and has a range from zero to six. Finally, Maximum ADSI is an estimate (larger than ADSI) of the largest code size. The value of the difference between ADSI and MADSI is used by COCOMOID as the primary code size risk parameter.

Text 14 - Maintenance sizing and development costs

The second part of the sizing section, Text 14, is for annual change traffic (ACT) and the standard monthly burdened cost for programmers, etc. This is the right half of the previous table. The ACT value is the primary determinate of code size that must be maintained during maintenance. It uses ADSI (since all code must be maintained) as its baseline. The standard monthly burdened cost for "programmers Cost/Mo" input is straight COCOMO. COCOMOID provides three other alternatives to cost effort by in addition to this simple standard; SCOW, by activity, and by phase in detailed model.

The fourth and last section we will discuss is the attribute inputs, Text 15 and Text 16. Some inputs are a single entry if it is not used for the COCOMO maintenance model. These have a "." after the variable name to indicate they are used only during

[22] Design Modification, Code Modification, and Integration Modification are the three pieces used by the COCOMO reuse model. ADSI values are adjusted by these and their percentage weights to determine how much equivalent code must be produced. The formula is in this case: (.4 * DM) + (.3 * CM) + (.3 * IM) = EDSI, Equivalent Delivered Source Instructions. These percentages are the general standard for COCOMO, but should be changed for other modeling conditions.

[23] [SEE], pg 538

development. The " ' " after the variable name is a reminder this is a module level attribute

```
Rg0-14 for M# --v
Component  IC   M RELY.DATA CPLX'TIME STOR VMVH TURN ACAP AEXP PCAP'VEXP'LEXP'
Mis E/F Proc1   1    5   55   55   33   55   33   33   33   33   33   33   33
...
```
Text 15 - First section of attributes inputs

and these inputs can vary within a subsystem. Subsystem attributes, without the "'", where the "M" column number is the same, must have the same inputs on attributes.

This numerical indicator for attribute inputs is to indicate the normal COCOMO word ratings as follows: 1 = Very Low, 2 = Low, 3 = Nominal, 4 = High, 5 = Very High, 6 = Extra High, and 7 = Extra Extra High. Since development and maintenance are both done in COCOMOID, dual entries are used. The left entry is for development and the right entry is for maintenance, e.g. "54" would mean a "Very High" rating for development and a "High" rating for maintenance.

The "XTRA" attribute in Text 16 will have no effect unless its data files are set up with new values instead of 1's in the password protected files they are located

```
Rg0-14 for M# --v                        RVOL for Maint
Component  IC   M MODP.TOOL SCED.SECU RUSE.VMVT XTRA
Mis E/F Proc1   1    3   44   3   33   33   33   33
...
```
Text 16 - Attributes, second part

in. It is fully available whenever it may be needed.

Let's summarize how to use key COCOMOID features. An important feature is using COCOMOID as an excellent training tool. The spreadsheet is set up close to the same format, see Text 17, as is used within [SEE] for the COCOMO CLEF[24], etc. formats and others. Since the formats is very similar and the calculations are the same, a student or someone new to COCOMO can more easily follow the COCOMO process and logic using [SEE].

---

[24] CLEF stands for the Component-Level Estimation Form used for Intermediate COCOMO.

DEVELOPMENT			Selected	(3)	(19)	(20)	MM	EDSI/	Cal MMi	-----------	$/
Component	IC	M	EDSI	AAF	EAF	MM NOM	DEV/ AM	MMorACT	Input	$	EDSI
INCR TEST	1	1	30000	100.0%	1.20	142.12	171.12	175.31		$889,844	30
INCR TEST	2	2	20000	100.0%	1.20	94.75	114.08	175.31		$593,229	30
INCR TEST	3	1	30000	100.0%	1.20	142.12	171.12	175.31		$889,844	30
			0	0	0.0%	0.47	0.00	0.00	NA		$0	NA
			0	0	0.0%	0.57	0.00	0.00	NA		$0	NA
			0	0	0.0%	0.72	0.00	0.00	NA		$0	NA
			0	0	0.0%	1.00	0.00	0.00	NA		$0	NA
			0	0	0.0%	1.00	0.00	0.00	NA		$0	NA
			0	0	0.0%	1.00	0.00	0.00	NA		$0	NA
			0	0	0.0%	1.00	0.00	0.00	NA		$0	NA
			0	0	0.0%	1.00	0.00	0.00	NA		$0	NA
			0	0	0.0%	1.00	0.00	0.00	NA		$0	NA

DEVELOPMENT			===========Eff====Eff%==	Adj	Sch	============Sched========$'s=====Max===					
Total EDSI			80000	PD	93.5	20.50%	36.50%	MYrs	2.5	486,448	80000
(MM)nom,EDSI/1000			379.0	DD	125.5	27.50%	40.00%	MHrs	4639	652,552	379.0
EDSI/MM, Nom Schd			211	CUT	112.9	24.75%		Months,T(d)	30.5	587,297	211
#Logic ERR=0			tables	IT	124.4	27.25%	23.50%	MDays	579.9	646,620	30
Pages 145-152			^Datafile	TOTAL	456.3	EDSI/MO	2621	Avg# Per	15.0	$2,372,918	

Text **17** - COCOMO CLEF output

The use of the global options allows very fast analysis of a great number of <u>reasonable alternates</u> with very little data entry. These include risk for model and size, quick "what-if's" from minimum to maximum sizes, saving and retrieving multiple data files, and alternative costing options and checks. The

$s in 1000s		Basic $	Int $	Detl $	#SDs=0	ERRs=0.00 03-Jan-91
Feasibility		0	0	0	Cal#=0	20.0% = 1 SD Rk
Req Analysis		253	237	540	Mode=Emb	0.0% Exp Sz Rk
Development, normal		2534	2373	5401	Vers=Prc	0.0% Max Sz Rk
Other		89	83	189	CAUTION:Incremental Dev Used
Maintenance		2209	1800	1800	Data: tables	Constrained
Total		5085	4493	7930	Detailed	Avail MMs/
EFFORT	MM-Limit		Man Months			Dev$ Avg Man ph,Manning
Plans&Req	0.0	48.7	45.6	103.9	540	7.2	0.00
Prod Dsgn	7.0	99.9	93.5	166.1	864	14.0	7.01
Program,DD	8.1	134.0	125.5	228.5	1188	16.2	8.09
CUT	8.2	120.6	112.9	231.8	1205	16.4	8.20
Intgr&Tst	14.5	132.8	124.4	412.4	2144	28.9	14.46
Total	12.0	536.0	502.0	1142.6	5941	20.9 Lim .5 avg^
SCHEDULE		Months			74% Constrained. has unadded
Plans&Req		10.4	11.0	14.5	14.5 SCED $ % Penalty of:
Prod Dsgn		7.3	9.0	11.9	23.7	0.0%:Manual Add
Program		8.7	10.7	14.1	28.2	0.0%PERT Sz Rk
Intgr&Tst		8.8	10.8	14.3	28.5	0.0%Rng Rk add
Total		35.2	41.5	54.7	95.0	0.0%Tot Rk add

Text **18** - COCOMOID's Comparison Outputs of COCOMO models

consolidation of effort, schedule, and cost results on one output screen, such as Text 18, allows fast visual views, in text and graphical form, of the results for all options chosen. Notations to document the estimate are integrated to data files and are a fast way to enter key exception information as well. Finally, incremental development is integrated in with other inputs - only the additional new required inputs need be entered. And the layout for incremental development is the same as described in Dr Boehm's paper on the first refinement to COCOMO.

There are several standard <u>outputs</u> to document an estimate. There are 18 <u>graphs</u>, and 11 printed <u>reports</u>, ranging from 2 to 8 pages each. Some sample COCOMOID outputs are shown in this paper. Text 18 is a selected portion from the main SUMMARY report. It contains cost, effort, and schedule for the basic, intermediate, and detailed results. All results are adjusted by options selected. The SUMMARY report is the only place risk adjustments are done. All other outputs that use formats similar to COCOMO are unadjusted so direct comparisons and training with SEE can be done. Text 17 is the COCOMO CLEF outputs matching closely to the book format. It is used in the SUMMARY and DEVELOPMENT DETAIL reports of COCOMOID. The individual attribute values are in a separate section and are not shown here.

The eleven available reports are: 1) Summary, 2) Development detail, 3) Maintenance detail, 4) COCOMO detailed model, 5) User documentation notes, 6) Calibration information, 7) Other factored costs, 8) Incremental development detail, 9) CLEF form summaries, 10) The entire worksheet, 11) All tables.

A sample graph, software manning by activity by phase, is shown in Figure 1. Another graphic sample is the incremental development graph showing the four phases for up to six increments in Figure 2. All graphic results are always quickly available for viewing from the COCOMOID menu. All can be printed out with any product working with PIC files.

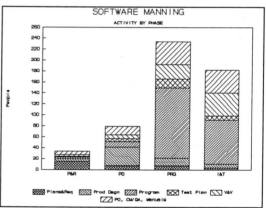

Figure 1 - Software Manning Activities by Phase

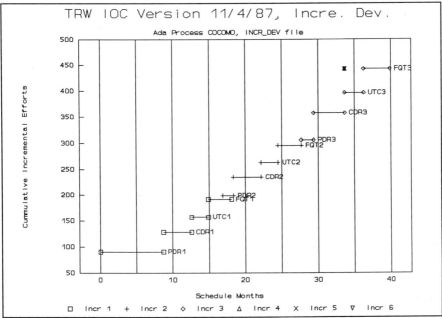

Figure 2 - Incremental Development Schedule and Effort for up to 6 Increments

The total set of 18 graphs are: 1) Life cycle total costs; 2) Development costs by phase using the detailed model; 3) Effort by phase for basic, intermediate, and detailed models; 4) Schedule by phase for basic, intermediate, detailed models, and constrained manpower; 5) Schedule constraint curve; 6) SCED adjustment effect; 7) Incremental development effort and schedule for up to six increments (Figure 2); 8) Development and maintenance for individual program entries; 9) Normal and constrained manning by phase with schedule; 10) Effort by phase for development; 11) SCOW monthly programmer rates; 12) SCOW selected rates compared with input rates; 13) Organic Rayleigh curve; 14) Intermediate model attributes; 15) Typical maintenance cost spread by activity; 16) Maintenance personnel equivalents and schedule by activity; 17) PERT results; and 18) Percent of eight functions performed in each of the four phases of development.

This concludes this introduction to the complete COCOMO model and COCOMOID. Distributing the COCOMOID family of worksheets is without license fee and remains the author property. Copies of COCOMOID are available to registering users by sending 360K formatted disk in preaddressed, prepaid mailer to either:

1) AFCSTC/ADT, ATTN: COCOMOID, 1111 Jeff-Davis Hwy, Cystal Gateway North, Suite 403, Arlington, VA 22202, AV 286-5869
2) SCEA, 101 S. Whiting Street, Suite 313, Alexandria, VA 22304, 703-751-8069.

Appendix A - Definitions from "Software Engineering Economics"

A. Development Mode Equations (Table 8-1, Page 117), original model

Development Mode   Nominal Effort Equation
Organic $(MM)nom = 3.2 * (KDSI)^{1.05}$
Semidetached $(MM)nom = 3.0 * (KDSI)^{1.12}$
Embedded $(MM)nom = 2.8 * (KDSI)^{1.20}$

B. Modes: Three Modes Are Used: Organic, Semidetached & Embedded. Below are the definitions for use of these modes.

1. Organic (Page 78): Relatively small software teams develop software in a highly familiar, in-house environment. This is a generally stable development environment, with very little concurrent development of associated new hardware and operational procedures. There is minimal need for innovative data processing architectures or algorithms, a relatively low premium on early completion of the project, and relatively small size. Very few organic-mode projects have developed products with more than 50 KDSI of new software. Larger organic-mode products often may be developed by using existing software. Teams generally have extensive experience working with related systems within the organization. And teams can

generally negotiate a modification of the specification that can be developed more easily and that will not be too difficult for the user to accommodate.

2. Semidetached (Page 79): An intermediate level of the project characteristic or a mixture of the organic and embedded mode characteristics. Thus, "experience in working with related software systems" could be characterized by; 1) team members all have intermediate level of experience with related systems, 2) team has a wide mixture of experienced and inexperienced people, or 3) team members have experience related to some aspects of development, but not others.

3. Embedded (Page 79): The major distinguishing factor is a need to operate within tight constraints. The product must operate within (is embedded in) a strongly coupled complex of hardware, software, regulations, and operational procedures, such as an air traffic control system. In general, the costs of changing the other parts of this complex are so high that their characteristics are considered essentially unchangeable, and the software is expected both to conform to their specifications, and to take up the slack on any unforeseen difficulties encountered or changes required within the other parts of the complex. As a result, the embedded-mode project does not generally have the option of negotiating easier software changes and fixes by modifying the requirements and interface specifications.

From a features perspective, the table below (Table 6-3, [SEE]) assists in determining which mode is the most suitable to use for COCOMO.

Models	Versions						Modes		
Original	Basic	Inter-mediate	Detailed	Incre-mental	Maint-enance	Cali-bration	Organic	Semi-detached	Embedded
Enhanced	x x								
Ada	x x / xx								
Ada Process	x x / x x								

Matrix Table of All Forms of COCOMO

MODE ATTRIBUTES TABLE

Feature	Organic	Semidetached	Embedded
Organizational understanding of product objectives	Thorough	Considerable	General
Experience in working with related software systems	Extensive	Considerable	Moderate
Need for software conformance with pre-established requirements	Basic	Considerable	Full
Need for software conformance with external interface specifications	Basic	Considerable	Full
Concurrent development of associated new hardware & operational procedures	Some	Moderate	Extensive
Need for innovative data process architectures, algorithms	Minimal	Some	Considerable
Premium on early completion	Low	Medium	High
Product size range	<50 KDSI	<300 KDSI	All Sizes
Example: Larger list in [SEE]	Batch data reduction	Most transaction process system	Avionics, Large,complex, transaction process

Appendix B - COCOMO Attribute Values

COCOMO Intermediate **Ada ProcessModel Attribute Values** with Interpolated ACAP, PCAP values (they change according to actual inputs used, this is a sample and matches Figure 3). The requirements volatility (RVOL) values are shown in the table, but not the graphs, since they are not used for development in COCOMO.

	RELY	DATA	CPLX'	TIME	STOR	VMVH	TURN	ACAP	AEXP	PCAP'	VEXP'	LEXP'	MODP	TOOL	SCED	SECU	RUSE	VMVT	XTRA	RqVol
VL	0.75	0.94	0.73	1	1	0.92	0.79	1.57	1.29	1.3	1.21	1.26	1.24	1.24	1.23	1	1	0.93	1	1
Low	0.88	0.94	0.85	1	1	0.92	0.87	1.29	1.13	1.12	1.1	1.14	1.1	1.1	1.08	1	1	0.93	1	0.91
Nom	0.96	1	0.97	1	1	1	1	1	1	1	1.04	0.98	1	1	1	1	1	1	1	0.91
High	1.07	1.08	1.08	1.11	1.06	1.09	1.07	0.8	0.91	0.89	0.9	0.95	0.86	0.91	1	1.1	1.1	1.07	1	1
VH	1.24	1.16	1.22	1.3	1.21	1.17	1.15	0.61	0.82	0.8	0.9	0.86	0.78	0.83	1	1.1	1.3	1.16	1	1.19
EH	1.24	1.16	1.43	1.66	1.56	1.17	1.15	0.61	0.82	0.8	0.9	0.86	0.78	0.73	1	1.1	1.5	1.16	1	1.38
XH	1.24	1.16	1.43	1.66	1.56	1.17	1.15	0.61	0.82	0.8	0.9	0.86	0.78	0.62	1	1.1	1.5	1.16	1	1.62

COCOMO Attribute Listing and Where Used

Category	Name	Description	Ori	Enh	Ada	Prc	Mnt	Cal
Product	RELY.	Required Software Reliability	Y	Y	C	C	N	Y
	DATA	Data Base Size	Y	Y	Y	Y	Y	Y
	CPLX'	Product Complexity	Y	Y	C	C	Y	Y
	RUSE.	Required Reusability	N	Y	Y	Y	C	C
Computer	TIME	Execution Time Constraint	Y	Y	Y	Y	Y	Y
	STOR	Main Storage Constraint	Y	Y	Y	Y	Y	Y
	VIRT	Virtual Machine Volatility	Y	N	N	N	N	Y
	VMVH	Host - Virtual Machine Volatility	N	Y	Y	Y	Y	Y
	VMVT	Target - Virtual Machine Volatility	N	Y	Y	Y	Y	Y
	TURN	Computer Turnaround Time	Y	C	Y	Y	Y	Y
Personnel	ACAP	Analyst Capability	Y	Y	Y	C	Y	Y
	AEXP	Applications Experience	Y	Y	Y	Y	Y	Y
	PCAP'	Programmer Capability	Y	Y	Y	C	Y	Y
	VEXP'	Virtual Machine Experience	Y	Y	Y	Y	Y	Y
	LEXP'	Programming Language Experience	Y	Y	C	C	Y	Y
Project	MODP.	Modern Programming Practices	Y	Y	Y	C	Y	Y
	TOOL	Use of Software Tools	Y	C	Y	Y	Y	Y
	SCED.	Required Development Schedule	Y	C	Y	Y	Y	Y
	SECU	Security	N	Y	Y	Y	Y	Y
	RVOL	Requirements Volatility	N	N	N	N	Y	Y
	XTRA	Extra	N	N	N	N	N	N

Definitions of above abbreviations:

Ori = Original
Enh = Enhanced
Ada = Ada
Prc = Ada Process
Mnt = Maintenance
Cal = Calibration

Graphic representations of the four basic Intermediate COCOMO Attribute Values are:

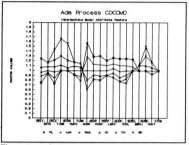

Figure 3 - Ada Process

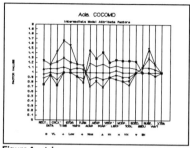

Figure 4 - Ada

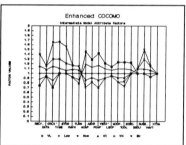

Figure 5 - Enhanced

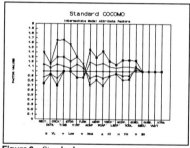

Figure 6 - Standard

Appendix C - Σ rating scale aids for determining Σ ratings at early stages. The overall chart is followed by more detail ones. These show typical characteristics of rating scale levels for the following COCOMO embedded Ada process formula::

$$MM_{NOM} = (2.8)(KDSI)^{1.04 + \sum_{i=1}^{4} W_i}$$

Weights W_i	0.00	0.01	0.02	0.03	0.04	0.05
Experience with the Ada process model	Successful on >1 mission-critical project	Successful on 1 mission-critical project	General familiarity with practices	Some familiarity with practices	Little familiarity with practices	No familiarity with practices
Design thoroughness at PDR: Unit package specs compiled, bodies outlined	Fully (100%)	Mostly (90%)	Generally (75%)	Often (60%)	Some (40%)	Little (20%)
Risks eliminated by PDR	Fully (100%)	Mostly (90%)	Generally (75%)	Often (60%)	Some (40%)	Little (20%)
Requirements volatility during development	No changes	Small noncritical changes	Frequent noncritical changes	Occasional moderate changes	Frequent moderate changes	Many large changes

Appendix C - Ada COCOMO Σ Factor: Design thoroughness by PDR

Characteristic	0.00	0.01	0.02	0.03	0.04	0.05
Schedule, budget, and internal milestones through PDR compatible with risk management plan	Fully	Mostly	Generally	Some	Little	None
Percent of development schedule devoted to preliminary design phase	40	33	25	17	10	5
Percent of required top software architects available to project	120	100	80	60	40	20
Tool support for developing and verifying Ada package specs	Full	Strong	Good	Some	Little	None
Level of uncertainty in key architecture drivers: mission, user interface, hardware, COTS, technology, performance	Very little	Little	Some	Considerable	Significant	Extreme

Appendix C - Ada COCOMO Σ Factor: Risk elimination by PDR

Characteristic	0.00	0.01	0.02	0.03	0.04	0.05
Risk management plan identifies all critical risk items, establishes milestones for resolving them by PDR	Fully	Mostly	Generally	Some	Little	None
Schedule, budget, and internal milestones through PDR compatible with risk management plan	Fully	Mostly	Generally	Some	Little	None
Percent of development schedule devoted to preliminary design phase	40	33	25	17	10	5
Percent of required top software architects available to project	120	100	80	60	40	20
Tool support available for resolving risk items	Full	Strong	Good	Some	Little	None
Number and criticality of risk items	< = 5, noncritical	>5, noncritical	1, noncritical	2-4, critical	5-10, critical	>10, critical

Appendix C - Ada COCOMO Σ factor: Requirements volatility.

Characteristic	0.00	0.01	0.02	0.03	0.04	0.05
System requirements baselined under rigorous change control	Fully	Mostly	Generally	Some	Little	None
Level of uncertainty in key requirements areas, mission, user interface, hardware, other interfaces	Very little	Little	Some	Considerable	Significant	Extreme
Organizational track record in keeping requirements stable	Excellent	Strong	Good	Moderate	Weak	Very weak
Used incremental development to stabilize requirements	Full	Strong	Good	Some	Little	None
System architecture modularized around major sources of change	Fully	Mostly	Generally	Some	Little	None

Ada Process COCOMO Σ factor: Complementary Maintenance Effects

Use of MPPs in development, previous maintenance	0.00 Routine	0.03 General	0.06 Often	0.09 Some	0.12 Little	0.15 None
Maintenance conformance to Ada process model	0.00 Full	0.01 General	0.02 Often	0.03 Some	0.04 Little	0.05 None

Appendix D - Tables for ACAP, PCAP, Effort %, and Schedule % interpolation.

Tables for ACAP, PCAP

Cost Driver	VERY LOW	LOW	NOMINAL	HIGH	VERY HIGH
ACAP	1.57-1.46	1.29-1.19	1.0-1.0	0.80-0.86	0.61-071
PCAP	1.3-1.42	1.12-1.17	1.0-1.0	0.89-0.86	0.80-0.70

Tables for Effort %, and Schedule %

Effort Distribution	SIZE				
Phase	Small, 2KDSI	Intermediate, 8 KDSI	Medium, 32 KDSI	Large, 128 KDSI	Very Large, 512 KDSI
Plans and Requirements	12-8	12-8	12-8	12-8	12-8
Product Design	23-18	23-18	23-18	23-18	23-18
Programming	57-60	55-57	53-54	51-51	49-48
Detailed Design	32-28	31-27	30-26	29-25	28-24
Code and Unit Test	25-32	24-30	23-28	22-26	21-24
Integration and Test	20-22	22-25	24-28	26-31	28-34
Schedule Distribution					
Plans and Requirements	28-24	32-28	36-32	40-36	44-40
Product Design	33-30	35-32	37-34	39-36	41-38
Programming	52-48	48-44	44-40	40-36	36-32
Integration and Test	15-22	17-24	19-26	21-28	23-30

Appendix E - COCOMOID Input Forms

COCOMOID, V3.20, GLOBAL INPUTS FORM

```
COCOMOID 3.200    Extra <RETURN> stops Global input    DATE->  __-__-_

    Data only name>        _____ <Proj 2,ID 4,What-ifs 2    0 LOGIC ERR's

                                                         ____ <Mnhr/Mo152

COCOMO Version >        _  0,Bk;1,Enh;2,Ada;3,Proc       ____ <Mndys/Mo19

Development Mode>       _  1=Org,2=Semi,3=Embedded

SEE Ada detail:    Exper/MPP Dsn/Cnf Risks. Req Vol. $_____Feasibility

Ada process, 0-5>      _                              Alt Mo $ Activity

  Calibration 0  >     _  0=Std,1-14=input table|      _____ Req Anal

  LOC Opt In     >     _  0=In,1=Min,2=Exp,3=Max|      _____ Prod Dsg

  Model Risk Incl>     _  0-3, # * 20%                 _____ Progm

SCOW $ Use = 1  >      _      _ <Add PERT Risk         _____ Test Pln

Programmer SCOW   Starting Median Top    Base Yr       _____ V&V

    Salary range>  $_,___ $_,___ $_,___    ____        _____ Proj Off

Salary SCOW          O/H    G&A  Profit Infl BY        _____ CM/QA

  Loading Factors>    _._%   _._%   _._%   _._%         _____ Manuals

PERT Min %    >      _% 0%-50% Range, 25% Rmd          _._%Overtime%

                    a Rmd 3,m Rmd 2,b SDs,0-3          % Mnt Profile

PERT Weights  >      1     _     _     _               ___.% 1st yr

                   # Yrs   $ Diff% IMEM.4-2.5,Pg540    ___.% 2nd yr

Maintenance   >      _    _%   _._                     ___.% 3rd yr
```

COCOMOID CODE-VALUES INPUT FORM

Mod Spread -->		Adapted	0.4	0.3	0.3	Cnvs Pln		Anl Chg	
	>	Del Src	Design	Code	Intg	Inc,Pg558	Max> ADSI	Traffic	Programmer
Rg0-14 for M# --v		Instruction		(0-100)	(0-200)	(0-6)	(#>ADSI)	(0-49)	Cost/Mo
Component IC	M	ADSI	DM	Reuse,CM	IM	CPI	MADSI	ACT	

Component										
_____	_	1 _____	__	__	__		_ _____	_____%	$____	
_____	-	_ _____	__	__	__		_ _____	_____%	$____	
_____	-	_ _____	__	__	__		_ _____	_____%	$____	
_____	-	_ _____	__	__	__		_ _____	_____%	$____	
_____	-	_ _____	__	__	__		_ _____	_____%	$____	
_____	-	_ _____	__	__	__		_ _____	_____%	$____	
_____	-	_ _____	__	__	__		_ _____	_____%	$____	
_____	-	_ _____	__	__	__		_ _____	_____%	$____	
_____	-	_ _____	__	__	__		_ _____	_____%	$____	
_____	-	_ _____	__	__	__		_ _____	_____%	$____	
_____	-	_ _____	__	__	__		_ _____	_____%	$____	
_____	-	_ _____	__	__	__		_ _____	_____%	$____	
_____	-	_ _____	__	__	__		_ _____	_____%	$____	
_____	-	_ _____	__	__	__		_ _____	_____%	$____	

Rg0-14 for M# --v COCOMOID ATTRIBUTES INPUT FORM   RVOL for Maint
  Component   M RELY.DATA CPLX'TIME STOR VIRT TURN ACAP AEXP PCAP'VEXP'LEXP'MODP.TOOL SCED.SECU RUSE.VMVT XTRA

_____ - _ _ _ _ _ _ _ _ _ _ _ _ _ _ _ _ _ _ _ _
_____ - _ _ _ _ _ _ _ _ _ _ _ _ _ _ _ _ _ _ _ _
_____ - _ _ _ _ _ _ _ _ _ _ _ _ _ _ _ _ _ _ _ _
_____ - _ _ _ _ _ _ _ _ _ _ _ _ _ _ _ _ _ _ _ _
_____ - _ _ _ _ _ _ _ _ _ _ _ _ _ _ _ _ _ _ _ _
_____ - _ _ _ _ _ _ _ _ _ _ _ _ _ _ _ _ _ _ _ _
_____ - _ _ _ _ _ _ _ _ _ _ _ _ _ _ _ _ _ _ _ _
_____ - _ _ _ _ _ _ _ _ _ _ _ _ _ _ _ _ _ _ _ _
_____ - _ _ _ _ _ _ _ _ _ _ _ _ _ _ _ _ _ _ _ _
_____ - _ _ _ _ _ _ _ _ _ _ _ _ _ _ _ _ _ _ _ _
_____ - _ _ _ _ _ _ _ _ _ _ _ _ _ _ _ _ _ _ _ _
_____ - _ _ _ _ _ _ _ _ _ _ _ _ _ _ _ _ _ _ _ _
_____ - _ _ _ _ _ _ _ _ _ _ _ _ _ _ _ _ _ _ _ _
_____ - _ _ _ _ _ _ _ _ _ _ _ _ _ _ _ _ _ _ _ _

BIBLIOGRAPHY

Boehm, Barry W. Software Engineering Economics. Englewood Cliffs, N. J.: Prentice-Hall, Inc., 1981.

Boehm, Barry W. and Royce, Walker, Ada(TM) COCOMO: TRW IOC VERSION, Proceeding, Third COCOMO Users' Group Meeting CMU Software Engineering Institute, Pittsburgh, PA

Fairley, Richard "Software Engineering Concepts" NY, NY: McGraw-Hill, 1985.

Kile, Raymond L. REVIC Software Cost Estimating Model User's Manual Version 8.7. New Mexico: Kirtland AFB, 1989.

Software Engineering Institute, The Fourth International COCOMO Users' Group Meeting, Oct 1988, Pittsburgh, Pennsylvania, Carnegie Mellon Univeristy.

Software Engineering Institute, The Fifth International COCOMO Users' Group Meeting, 17-19 Oct 1989, Pittsburgh, Pennsylvania, Carnegie Mellon University.

U.S. Department of the Air Force. Revic Software Cost Estimating Model User's Manual Version 8.7 by Raymond L. Kile, U.S. Department of the Air Force, Kirtland Air Force Base, New Mexico, 1989.

Vick, C. R., and C. V. Ramamoorthy, ed. Handbook of Software Engineering. New York: Van Nostrand Reinhold Company, 1984.

III. Force Costing

Calculating the Cost Savings From Reducing Military Force Structure: How to Formulate the Cost Problem

Michael G. Shanley

RAND, 1700 Main Street, Santa Monica, CA 90407

INTRODUCTION

World events, such as the Gulf War and the breakup of the Soviet Union and changes in NATO, have led to dramatic changes in the defense environment and in the associated requirements for an adequate defense. These changes have motivated a far reaching reconsideration of U.S. military force structure that will lead to a major restructuring of military forces within the next few years. Proposed changes include: a major reduction in force size (i.e., a reduction in the number of military units, such as infantry battalions, aircraft squadrons, or ships of various types) along with a corresponding reorganization of force-wide support organizations and resource management systems; a realignment of the remaining units on existing bases, closing those bases that are no longer required; a reorganization of units to smaller, more mobile and flexible entities; and a change in the mix of active and reserve units. Plans to reduce the defense budget rely heavily on such changes in force structure.

Planning the transition to the new force structure requires speedy and accurate cost analysis. A large number of both large and small force structure changes will be proposed, with complex and widely differing impacts in costs. Not all would save money, not even those that reduce overall force size. Moreover, the magnitude of the cost impacts will often not be intuitive, and most proposed changes would be expensive and time consuming to modify once implementation was under way. Thus, decision makers require timely and accurate information about the cost impacts of a large number of alternatives.

Unfortunately, current technology for estimating the cost impact of force structure change does not satisfy the requirements of decision makers. Competing cost estimates of a single force structure option often vary widely, even for small, routine changes. Because an insufficient basis is typically provided to explain the differences[1] among competing estimates, decisionmakers are essentially left to guess which figure is correct or to decide on a "compromise" amount.

This article argues that a major explanation for the disparity in competing force structure cost estimates lies in differences in *how problems are defined*. To solve this problem, this article proposes a formal set of guidelines to support the problem formulation stage of the force structure cost analysis. Below, we first elaborate on the nature of the costing problem in

---

[1] Sometimes studies lack sufficient documentation for comparing with the outcomes with other studies. In other cases, the approach taken by two studies is so different that competing estimates cannot be reconciled.

this area, then develop the set of guidelines that, if implemented, would greatly increase the accuracy of force structure cost estimation. We conclude with a discussion of how a similar tactics could be applied to other policy arenas.

THE FORCE STRUCTURE COSTING PROBLEM

The assessment of the cost implications of force structure change occurs regularly in the context of the Planning, Programming, and Budgeting System (PPBS), the formal mechanism used by the Department of Defense to manage the allocation of defense resources and to formulate proposed budgets. Issues of force structure change arise first with the Services themselves, when they develop their future year defense plan (FYDP). Force structure issues again arise in the review of those plans by the Office of the Secretary of Defense (OSD) during the programming and budgeting cycles of the PPBS.

In both the Service and OSD contexts, cost analysts are often asked to evaluate the gross budget implications of an incremental change to the existing force structure. A variety of organizations (or offices within the same organization) may address the same problem, under the direction of single (or at most a few) analysts with varying expertise in force structure costing, and using a variety of methodologies and data sources. Descriptions of the the the proposed change are often left vague, with the analyst left to fill in the missing details. For example, cost analysts may have to work with only the specification of a generic type of unit (e.g., an Air Force F-15 squadron or an Army mechinized infantry division), and a common description of the type of change units would undergo (e.g., a deactivation or transfer to the Reserve components). Further, because a large and growing number of force structure changes are considered within each PPBS cycle, results on any one alternative are often needed in a hurry, sometimes within only days or hours of the original tasking.[2]

Table 1 provides three recent examples of the highly variable cost estimates that can arise from current force structure costing practices. Each of these particular cases involve relatively small scale active-to reserve-transfers considered in the context of the formulation of recent defense budgets. Two come from the Air Force, and one from the Navy. The Navy issue arose during the new Bush administration's review of the Reagan budget, and concerned the proposed transfer of 24 older frigates to the Reserve forces between 1990 and 1994. The C-5 case arose during development of the FY88-92 Air Force program, and involved the proposed transfer of 26 cargo aircraft to the Reserve forces.

Even though all three changes were proposed to save money, the multiple estimates by various analysts revealed huge differences in the 5 year savings impact. In fact, the range of estimates presented in the table documents a significant disagreement about not only *how much* would be saved, but also about *whether* there would be any savings at all.

---

[2] While we have described the cost environment addressed in this article, it is not the only type of force structure costing that occurs within the DOD. For example, developing a total budget from an agreed upon force structure involves a whole different set of tasks than those involved with evaluating alternative force structures. Costing for budget development requires considerably more detailed calculations to associate proposed spending with budget line item level-of-detail. In contrast to the scenario described above, this process requires the efforts of a large number of personnel over a considerable time period, the use and manipulation of large data bases and models, and frequent coordination among the many offices within the Services engaged in budget development.

Table 1

RANGE OF ESTIMATED SAVINGS FROM TRANSFERRING
ACTIVE MISSIONS TO THE RESERVE FORCES:
THREE EXAMPLES

	Program	Estimated 5 year Savings ($M)	
Transfers	Years	High	Low
24 FF-1052 ships	FY 90-94	550	-100
26 C-5 aircraft	FY 88-92	160	-300
1 KC-135 squadron	FY 90-94	150	-200

Estimated costs of force structure change reach such extremes largely because typical descriptions of proposed changes contain insufficient detail for costing purposes. Absent a well specified problem, cost analysts make a wide variety of assumptions about critical cost drivers, producing an large variance in results. In addition, because many analysts are not experts in force structure design, the assumptions made are not always appropriate; and even when they are appropriate, they are often not apparent in the presentation of results. As a result, policy makers receive inadequate guidance for decision making.

Short-hand descriptions of force structure change are not specific enough to capture the inherent complexities of the process that produces force structure change, complexities that also can have large impacts on costs. These complexities include the following:

- Changing one set of units can have indirect effects on other parts of the force. Indirect effects can occur because the units designated for a change share a peacetime or wartime mission (i.e., tasking), not mentioned in the original problem description. For example, if a proposal calls for the contraction of force structure but mission needs remain unchanged, other units may have to take up the slack. Similarly, units can be affected because they benefit from resources (e.g., manpower or equipment) released by units scheduled to contract, or because they lose resources to units scheduled to expand. Further, combat support units (e.g., maintenance or transportation units) can be indirectly affected by the units they support, expanding or contracting operations in response to a change in a named unit. If a proposed change is large enough, central support organizations (e.g., training and medical) that support the service as a whole may be similarly affected. Identifying the kind of indirect impacts mentioned above is often left to the cost analyst. Determining all the parts of the force that are affected by a change, and the net changes in resources expended, represents a major task.

- Units with the same label can vary significantly in terms of variables that drive costs, such as the amount of equipment, the number and composition of personnel, and the level of operation they maintain in peacetime. For example a "B-52 squadron" may actually have 13, 14, 15, 16, or 19 aircraft.[3] The number could even be smaller for a squadron in the Reserves. The same is true of generic descriptions of the type of force structure change. For example, as we will review in detail below, "transferring a unit from the actives to reserves" can have a wide variety of meanings, all with very different implications for cost and the capability of the resulting force. Again, it is typically left to the analyst to go below the generic level of "type of unit" and "type of change" to that of resources, activities, and missions.

---

[3] *Air Force Magazine*, May 1990, p. 50.

• The costs of transition, the expenses incurred during the change-over period from one force structure to another, can vary widely. These include the one-time, or non-recurring, costs of acquiring the manpower, equipment, and facilities needed to support the new force structure. Under some circumstances (an example will appear below), they also include the operating and support costs of some units during the changeover period. Transition costs tend to be unique to particular problems because there are many ways to implement a change. Their values vary widely even with a given type of change for a given set of units, often making the critical difference in a decision. This is especially important when short term savings is an important goal. Yet analysts often receive little or no information about how a proposed change might be implemented, and often have quite limited access to factors that might help make an estimate.

In summary, simple descriptions of proposed force structure changes usually ignore the indirect effect of the change and critical aspects of the implementation.

Some of the complexities of force structure change have been addressed in existing cost models. For example, many models require input distributed on a yearly basis (helping to capture the correct timing of transition costs), and most attempt to capture at least some of the effects that changes to combat units impose on support organizations. However, the use of models also contributes to the diversity of cost estimates because analysts do not agree on the most appropriate methodology to use. Cost models can lead to different results because some are more complete than others (e.g., some have limited consideration of transition costs), and because the models make widely differing assumptions about important cost driving factors (e.g., how much of the cost of the support base is fixed with regard a particular force structure change). Further, the assumptions made are not necessarily transparent to either the analyst or the decision maker.

Force structure costing also occurs in a political environment that creates certain incentives that skew the results of cost studies. Different cost estimates of a given force structure change are often produced by advocates of opposing positions who are willing, at a minimum, to resolve problem uncertainties in a way that favors their own position. For example, in studies of active-to-reserve transfers, groups in favor of the change may emphasize the potential operating and support cost savings, while those against the change may emphasize the cost of indirect effects on other units.

Incentives contrary to impartial cost analyses can also develop when analysts participate in budget cutting drills. The vary objective itself--taking money out of the budget--leads to less consideration of aspects of force structure changes that add to costs (e.g., the costs of the transition to a new force structure) than those that decrease cost. Further, when analysts and force structure planners operate under time pressure, they can acquire a vested interest in potential force structure changes identified early in the exercise, simply because inadequate time would be available to develop and analyze new alternatives. Rather than seeking new alternatives, favored force structure changes may be under-resourced (e.g., no funds provided to cover transition cost), thus, in effect, hiding the true cost of the change.

Cost estimates can also vary because analysts sometimes lack the knowledge or experience required to adequately address the problem. The military practice of rotating personnel to new jobs every few years leads to a constant supply of new analysts who, while they may be experienced in some aspects of cost analysis, may not be fully familiar with the complexities of force structure costing. Moreover, analysts often only minimally benefit from the experience of others, because cost studies rarely contained adequate resources to document how they are completed. In some cases, the failure of new analysts to accurately calculate costs may derive from an ignorance of some of the complexity or an unawareness of some aspect of implementation.

More often, however, shortcomings in cost studies are a matter of timing; for example, analysts are unable to find the right data or determine the most appropriate assumption or methodology to use within the time frame of the study. This is a common outcome because results in many cost studies are required in an extremely short period of time.

THE SOLUTION: GUIDELINES FOR PROBLEM FORMULATION

To empirically address the issue of widely varying cost estimates for single force structure changes, we undertook an analysis of the existing cost analysis process. Our case studies followed a number of force structure cost analyses that occurred in the PPBS process, from their initial stages to final results. While we found a variety of explanations for highly variable cost results (e.g., the use of a wide range of methodologies and data sources without adequate consideration of their deficiencies), the greatest extremes were produced by varying definitions of (what was supposed to be) the same problem.

In response, we developed the "guidelines" detailed in this article. The guidelines are designed as a planning tool for translating vaguely worded force structure proposals into ones that contain adequate detail for costing. In developing the guidelines, we sought to simulate the process followed by the experts--namely to extract and highlight critical information by asking a series of questions specific to the force structure policy area and the PPBS costing process. The major challenge in this endeavor was defining a list of questions that covered all the possibilities but did not become too detailed. On the one hand, a short list of general questions might fail to provide sufficient information for analysts to understand the issues sufficiently to apply the guidelines to their own work. On the other hand, a long list of questions that attempted to cover every possible complexity and contingency of force structure costing would become too cumbersome to use and would defeat the purpose of using the guidelines as a planning tool. We eventually designed a relatively short list of questions (15 major questions, most with secondary supporting questions), but one in which each question was supplemented by a series of examples. The examples serve not only to illustrate the importance of the questions and the complexity of some of the answers, but also act as a primer for analysts interested in developing the judgement required to apply the guidelines to their own work. Further, the use of examples was intended to encourage the development of case notebooks that analysts could build upon based on their own experience.

Although concrete examples are an integral part of the guidelines, space limitations make it difficult to adequately develop many in a short article. Thus, for purposes of the article, we focus on four case studies to use as examples. The case studies are first described comprehensively below, allowing a short-hand reference when discussed in the context of the guidelines themselves. Following the case study summaries, each of the 15 guidelines are reviewed individually to illustrate their role in proper problem formulation. Finally, to show how the guidelines might be taken to an additional level of detail, a methodology is explored for assisting the analyst in answering one of the questions posed.

Case Studies

The four major case studies used in this article are based on published cost analyses[4] dealing with changes in the mix of active and reserve units.[5] To provide an illustration of the importance of the guidelines, we first describe the cases in terms of how they would ordinarily be presented to cost analysts, then show the dollar difference of making different assumptions about the missing problem elements.

As typically-worded changes in the force structure, the four cases might be described to cost analysts as follows:

- Transfer of two C-141 squadrons from the active Air Force to the Air Reserve Forces.

- Transfer of 26 C-5As in the active Air Force to the Air Reserve Forces.

- Transfer of 24 FF-1052 ships from the active Navy to the Navy Reserve.

- Modernization of 15 AH-1 helicopter squadrons in the Army National Guard with AH-64s

The discussion below illustrates the point that varying the implementation leads to a large variation in the potential outcome.[6]

Transfer of C-141 Squadrons: This proposal specifies that two squadrons would transfer from the active Air Force to the Air Reserve Forces. If the change involves the deactivation of C-141 squadrons in the active forces and the modernization of existing squadrons in the reserves (whose old C-130 equipment would be mothballed), the changes could save a large amount--an estimated $151 million in operating and support costs per year in the steady state. Alternatively, if the transfers involve the establishment of new C-141 units in the reserves (rather than the modernization of existing units), the transfers would save less--about $93 million per year. However, suppose the external peacetime airlift and supplementary training duties of the deactivating units were considered important enough to maintain. Because the new Reserve squadrons could not undertake those functions with a part-time force, the job could fall to the remaining C-141 squadrons in the active force. The increase in their operation would reduce the savings still further--to $20 million per year. But even that savings would not materialize under some circumstances, such as if the personnel of the active units were simply reassigned and the total size of the active forces was not reduced to reflect the reduction in force structure. In that case, the transfer of C-141s to the reserves would actually *cost* money--about $40 million per year.

Not surprisingly, the effects on military capability vary inversely to the cost consequences. The alternative that saves the most involves a reduction of the force structure--

---

[4] The examples are derived from the following publications: Glenn A. Gotz, et al., *Estimating The Costs of Changes in the Active/Reserve Balance,* R-3748-FMP/PAE/JCS, Section II. and Michael Shanley, *Active/Reserve Cost Methodology: Case Studies* (R-3748/2-FMP/PAE).

[5] Several characteristics of the case studies limit their scope, although not the findings themselves. First, rather than focusing on all types of force structure change, the case studies are directed toward changes in the active/reserve mix of force units. Second, because most of the research was conducted in 1989, before proposals emerged for major force structure change, the case studies include relatively small changes in force structure. However, in retrospect, we believe we have created a tool that applies more generally to a wide range of force structure problems. Further, recent experience has shown that even when major changes in force structure are contemplated, they are often addressed in small segments, one at a time.

[6] Cost figures in the case studies come from the published documents referenced above. No attempt has been made to evaluate or update the information in those other documents; rather they have simply been used as a convenient source for illustrating the importance of proper problem formulation in the force structure context.

by two C-130 squadrons. The next most costly alternative avoids that reduction, keeping the C-130 missions. Saving even less, the third alternative also avoids the loss of the peacetime airlift services that would ordinarily accompany the transfers.[7] And the alternative that actually adds to cost also has an added benefit: more personnel to fill positions in the remaining active units.

Transfer of C-5A Cargo Aircraft: This proposed change involves the transfer of the 26 remaining C-5As in the active Air Force to the Air Reserve Forces. The cost outcome of this case study revolves around two questions: what the base case is, and whether unused facilities already exist. If the alternative is to leave the aircraft in the active forces (as it was in the C-141 case above), the transfers could well *increase* total costs, at least for the programming period considered in PPBS process. Although operating and support costs would decrease by roughly $12 million per year, those savings could easily be swamped by the immediate costs of building runways, hangars, and related facilities to support the C-5s in the Reserve Forces. For example, locating the aircraft at a commercial airfield in the Air National Guard unit was estimated to generate $160 million in construction costs alone if that airfield had not previously supported the C-5. Much of that construction cost could be avoided however, if the aircraft could be located on an existing C-5 active base that could accommodate them. Even then, however, the required investment for acquiring and training reserve pilots and crew would keep the move from becoming a major money saver.

All the costs of transition could be justified, however, if the base case were different--if, for instance, the C-5s were transferred to make room for a new aircraft, like the new C-17 coming off the production line. In that case, the alternative to the C-5 transfer would be building facilities for the new planes. Under this scenario, transferring the C-5s to the Reserves would free space for the new model, avoiding large new investments in facilities and land--larger even than the cost of providing facilities for the C-5s at their new locations, and training additional Reserve pilots.

Transfer of FF-1052 Ships: This proposal involves the transfer of 24 FF-1052 frigates in the active Navy to the Navy Reserve. The transfer of the FF-1052 frigates could generate a modest 5 percent savings in operating and support costs, yielding about $1 million per year per ship, or about $24 million per year in total. However, those savings could be negated by the required additional expenditures of support organizations. If the transfer required movement of the ships to new homeports where reserve personnel could be recruited, the cost of pier projects at those ports could amount to $216 million (nearly $9 million per ship). Further, if the separate maintenance organizations that service reserve ships (called Ship Intermediate Maintenance Activities or SIMAs) had to expand their workforce, support equipment, and facilities to handle the new workload, the initial investment could involve as much as another $120 million. Together, supporting organizations could have to spend up to $340 million if applied to all 24 ships. How much of that would actually be required (anywhere from zero to $340 million) would depend on the extent of excess capacity in the supporting organizations. If only 10 percent of the transfers (i.e.,3 ships) required the investment, the transfers could begin saving money in less than 2 years, and the proposal could realize short term savings.

---

[7] Reserve units cannot provide the same airlift services as the same size active forces because of the parttime nature of reserve training.

Alternatively, if all ships required the investment, the time to breakeven (even ignoring discount rates) would increase to 14 years, well beyond the limit for a proposal aimed at near-term budget savings.

Modernization of AH-1 Squadrons with AH-64 Aircraft: This proposal involves the modernization of 15 AH-1 helicopter squadrons in the Air National Guard with new AH-64s. Normally, one would expect that transferring to newer, more capable equipment would require greater expenditures, but again, this depends on the definition of the base case. Suppose at least some of the AH-64s were coming from deactivating units in the active Army. In that case, the move could achieve considerable yearly savings, because AH-64 squadrons cost at least $7 million less to operate in the Guard than in the active Army. If, for example, a third of the Guard units received their helicopters from that source, the savings would amount to more than $35 million (5 x $7) per year after paying the modest costs of transition.

In contrast, operating costs would remain constant if the AH-64s were coming exclusively from the production line, and the alternative to modernizing the AH-1 squadrons in the Guard was to modernize similar squadrons in the active forces.[8] In this scenario, the comparison is no longer between operating the AH-64 helicopters in the actives versus operating those same helicopters in the Guard. Instead, the appropriate comparison is a) the difference between operation of an AH-64 squadron and an AH-1 squadron *in the Guard*, and b) the difference between operation of an AH-64 squadron and an AH-1 squadron *in the active forces*. As it turns out, those alternatives cost about the same amount, because while an AH-64 squadron is less expensive to operate in the Guard, so is an AH-1 squadron.

Even if savings from modernization in the Guard were possible, they could be delayed for many years if the AH-64 Guard units were required to maintain similar readiness during the transition periods as would occur in corresponding active modernizations. With their normal part-time schedule, Guard squadrons require about two and a half years to complete transition training from the AH-1 to AH-64 equipment, about one and a half years longer than it takes an active unit training full time. Maintaining extra AH-64 squadrons in the actives to cover for the Guard units during this additional train-up period would cost would cost about $38 million per squadron, eliminating the possibility of any short-term savings.

Guidelines

The examples above illustrate the consequences of overlooking important aspects of problem definition. They will also serve as illustrations for the 15 force structure guidelines discussed below. (The full set of guidelines, including the series of second-level questions that accompanies nearly all the first-level questions, appears in the appendix.) The guidelines are organized into three major subject areas: Force Structure Change, Changes During the Transition Period, and Net Changes in Resources, Activities, and Missions. Each question is listed, then explained and illustrated. Our intent here is not to be comprehensive in our discussion[9] but rather to demonstrate the value of a high-level, generic planning guide that is illustrated with examples.

---

[8]Note that the implicit assumption here is that the cost of producing the AH-64s is a sunk cost for purposes of evaluating where they should be placed in the force structure.

[9] Readers interested in comprehensiveness can refer to the original report, Michael Shanley, *Guidelines for Planning the Cost Analysis of Active/Reserve Force Structure Change*, R-4061-FMP/PAE.

Force Structure Change

The first series of questions extracts the exact change in force structure. To understand the force structure change, the analyst needs to establish a base case from which to calibrate the change. The analyst also needs to identify all units affected, directly or indirectly; the type of change (e.g., activation, deactivation, modernization) units are expected to undergo; and any changes to the units' supporting infrastructure. Finally, because proposed changes in force structure are often part of a more comprehensive cost-cutting plan, analysts need to isolate the effects of force structure changes from the effects of other measures to reduce costs.

1. What are the characteristics of the "base case"?

Because the goal of a force structure cost analysis is to calculate the *difference* between the cost of the new alternative and the base case, establishing the base case represents half of the equation for determining whether a change yields a savings or a cost. The C-5 and AH-64 examples above illustrated the dollar difference the choice of a base case can make. Other examples could easily have been constructed. For instance, in the FF-1052 study, moving the ships to the reserves saved resources only because the assumed alternative was to leave them in the active forces. If the base case alternative was instead to retire the ships, the move to the reserves would have represented a net *increase* in expenditure.

In addition to the use in the calculation of cost, base case identification can help focus analysis resources. For example, in the AH-64 case, knowing the base case not only identified what had to be calculated, but also what could be ignored. In particular, the example did not discuss the disposition of the replaced AH-1 helicopters in the modernized units, even though they might have been used to modernize other units with still older equipment. That information was irrelevant to the decision analysis because under both the base case (modernize active squadrons) and the proposed change (modernize Guard squadrons), the AH-1 aircraft would be released to the same locations.

2. What military units are affected, directly or indirectly, by this alternative?

Combat units named in a proposed force structure change can indirectly affect other deployable units by two basic means. First, units can affect others because of a shared mission, whereby changes specific to one unit disrupt the peacetime activity levels or wartime responsibilities of other units. This was true, for example, in one of the alternatives in the C-141 example, whereby non-specified cargo squadrons in the active forces compensated for the loss of airlift services brought about by the reserve transfer. A second way a force structure change can indirectly affect units is through a transfer of resources. For example, in the C-141 example, one alternative called for the transfer of the released personnel in the C-141 squadrons to other non-specified active units. When non-specified units are indirectly affected by a force structure change, they need to be added to the problem definition.

3. What is the "type" of each unit affected?

Proposed changes in force structure often deal with "types" of units, rather than specifically identified force units. In those cases "type of unit" is intended to provide enough description to generic units to permit the calculation of cost and the assessment of capability. To calculate costs, generic units *must* be described at least in terms of the military component (i.e., Active, Reserve, Guard); organizational level (e.g., squadron or wing, battalion or division), overall mission (e.g., infantry, tank), and when appropriate, the MDS (mission design series) of the weapon system. Other unit characteristics can also prove critical to cost.

For example, in the Air Force the "crew ratio" (defined as the number of aircrews per plane) determines the composition and cost of unit personnel and figures in the determination of equipment operating costs. The "unit status rating," used in all Services, indicates the extent a unit possesses all the required resources and has completed all the required training to undertake its full wartime mission. That information turned out to be important in the AH-64 case; the fact that the AH-1 squadron in the Guard had a lower rating than was programmed for the modernized AH-64 unit meant a much larger increase in resource use. While some of the information in unit descriptors might also be gathered by asking about changes in resources, activities, and missions (see Questions 11-15 below), unit descriptors provide an important independent source of information about unit cost, and a lead to and check on the accuracy of other information received.

4. What type of change will the affected units undergo?

To correctly calculate costs, problem statements must specify the precise type of change units are expected to undergo. Often simple descriptors (e.g., activation, deactivation, modernization, augmentation, transfer, reorganization) are inadequate. For example, consider the active to reserve "transfers" described in the examples above. On the active side, a decision to transfer could mean a reduction in the number of units (through deactivation) or a change in unit composition (through modernization or the reduction of unit equipment). On the reserve side (exemplified by the C-141 case), "transfer" could mean an increase in the number of units (through the creation of a new unit) or a change in unit composition (through modernization or the augmentation of unit equipment). Thus, analysts often need to probe for additional specifics to understand the precise nature of a change.

5. Does the number of units increase, decrease, or stay the same?

Accurate costing requires identifying changes in the number of units. For example, the case studies described above show that adding or subtracting units has a large impact on costs (see, in particular, the C-141 and AH-64 examples). In addition, even if major resource levels remain constant, the number of units is important to the calculation of cost because there may be economies or diseconomies of scale associated with different force structure configurations. For example, a recent proposal concerning Air Force structure called for consolidating all fighter aircraft into a smaller number of larger squadrons that could be based at fewer locations. That change was expected to realize economies of scale in the overhead costs of basing and managing those aircraft.[10]

6. Does the change affect the unit support structure?

Military combat units make demands on resources (e.g., manpower, equipment, supplies, and installations) and services (e.g., training, medical services, maintenance, supply, equipment production) supplied by support organizations. When analysts consider the cost consequences of force structure change, they must take into account not only the direct effect of the change on the combat units themselves, but also the indirect effects on the supporting infrastructure. In general, the larger the change in force structure, the greater likelihood the indirect effect on the support infrastructure will significantly affect the calculation of cost. In the FF-1052 case described above, it was the non-recurring costs of support organizations that turned out to be the critical factor.

---

[10] See *The Washington Post*, editorial page, Sunday, April 16, 1989.

Existing force structure models automatically capture some changes in support organizations (e.g., changes in the variable costs of repair depots), but most use over-simplifying assumptions (e.g., a given proportion of the costs are fixed or variable, or the support costs change in proportion to the change in the number of personnel) that are difficult to justify or even adjust in relation to specific cases of force structure change. Currently, if large changes in force structure indirectly affect the organization or management of support organizations, the cost analyst must independently identify and analyze the effects.

7. *Are other changes, unrelated to force structure, included in the problem definition?*

Proposals labeled as force structure changes can sometimes contain unrelated and separable strategies to alter the resourcing of units or their supporting infrastructure. One example is provided by the FF-1052 case. When the FF-1052 proposal was presented for approval during the PPBS process, it was labeled as a change in force structure; namely, the transfer of FF-1052 frigates. However, a careful reading of the proposed change showed that it contained a second cost-cutting measure (namely, a change in depot maintenance policy) that accounted for over three quarters of the savings impact. Proposed changes in force structure are combined with unit resourcing decisions in other ways as well: changes are proposed but full resourcing is withheld (e.g., new units are created but the total number of authorized personnel is left unchanged), or changes are proposed but some costs are simply ignored (e.g., force structure changes are required without budgeting for necessary transition activities). Cost analysts should distinguish force structure changes from unit resourcing changes because doing so fosters independent choices on separable decisions. For example, in the FF-1052 case, decision makers might have decided to change depot policy without making any change to force structure.

Changes During the Transition Period

A second set of questions addresses the transition tasks associated with the implementation of a proposed change, tasks which generate personnel, equipment and facilities related costs. The costs can occur not only due to changes in the units directly affected by the force structure change, but also due to changes in indirectly affected units and support organizations. Further, the analyst must search beyond the transition requirements to consider resources as well; some needs can be met by existing facilities and inventories. Finally, the timing of transition costs can also become an issue, especially when, as is often the case in a budgetary context, a primary interest is short-term savings.

8. *What are the transition tasks and the associated nonrecurring costs or savings?*

Transition costs or savings are those generated by the tasks required to implement a force structure change. The tasks include those for acquiring and processing personnel (generating, for example, relocation costs, enlistment bonuses, training costs, severance pay); purchasing and processing equipment (generating, for example, investment in support equipment or expenditures for moving major weapon systems); constructing or altering facilities and installations (generating, for example, costs for military construction); and planning and administering the change (generating, for example, costs for the design of a training program when a reserve unit receives a new weapon system). Transition savings are generated by the avoidance of cost, such as when the deactivation of a unit saves the cost of planned facility improvements. Case studies have shown that transition costs are often of sufficient magnitude to drive the outcomes of an analysis; in three of the four cases cited above, they represented a critical factor.

9. *To what extent can transition costs be saved by making use of untapped resources and excess capacity?*

The start-up costs of military units are driven by well-defined requirements for types of personnel, equipment, and facilities. However, non-recurring costs for a given type of force structure change are highly variable and difficult to compute (at least in the time frame of most studies) because transition costs also depend on the current supply of resources in inventory and current capacities in existing facilities and installations. The supply of existing facilities can be particularly difficult to compute because it critically depends on location, and specific installations are often not named in proposed force structure changes. For example, the transition costs involved with the transfer of 24 FF-1052s, as described above, ranged from zero to $340 million, depending on the availability of pier space and the capacities of existing maintenance organizations. As a result, any cost analysis of force structure change must address the potential that existing unused capacity can fulfill some of the requirements. Moreover, decision makers need to know how much the near-term cost implications depend on exactly how a force structure change is implemented.

10. *What role does timing play in costs?*

The timing of costs can play an important part in the outcome of a cost analysis. First, the timing of expenditures affects present value calculations; savings of a given size are more valuable in the short term than in the long term. Second, in the programming and budgeting context, timing often becomes an outcome as important as the cost itself. For example, in cost exercises designed to reduce expenses in the near term, force structure changes with significant long-term savings but a high initial transition cost are effectively disqualified. Third, how fast a force structure change is implemented can directly affect the cost. For example, the faster the pace of new unit formation, the greater is the likely per unit cost of procuring the unit's weapon system, and the greater the likely per person cost of acquiring the unit's personnel. Finally, overlapping cost can become a cost of transition. In the AH-64 case described above, a substantial transition cost was created by the need to overlap the operation of two units in order that mission capability could be maintained during the transition period.

Net Changes in Resources, Activities, and Missions

The third set of questions is intended to identify those changes in resource and activity levels that drive cost. To answer these questions, the analyst must first determine all the units affected by the change (see questions 2 and 6), then inquire directly into the *net* effect on manning type and quantity, equipment type and quantity, and on equipment operating tempo. The net effect of such changes does not necessarily follow directly from the nature of the force structure change. For example, unit deactivations are not always accompanied by decreases in total service personnel. Finally, analysts need to determine the capability implications of a proposed force structure change, so that after the costs (or savings) of the change can be calculated, decision makers can see what the dollars are buying.

11. *How does personnel endstrength change?*

"Personnel" refers to both the number and the type of personnel in units involved in a force structure change. Changes in personnel do not always parallel changes in force structure, because decisions about the total number of service personnel are independent from those about force structure. Thus, for example, unit deactivations may or may not be followed by reductions in overall personnel levels. The C-141 case illustrated the importance of that decision, where it meant the difference between a $20-million-per-year net *savings* and a $40-million-per-year net *cost*.

Even if the total number of personnel does change in parallel with force structure, personnel composition can have important implications for cost. For example, if a unit deactivation is accompanied by a decrease in new recruits rather than the more senior personnel in the deactivating unit, the savings will be significantly less (at least in the initial years) because new personnel cost less than experienced personnel. To take another example, if a new reserve fighter squadron must recruit and train new pilots, the costs can reach several million per pilot, far exceeding the cost of retraining pilots with prior service experience.

12. How does equipment change?

Unit equipment refers not only to a unit's major weapon system(s), but also to ground support equipment (for aviation units), maintenance support and test equipment, training equipment, other major items (e.g., trucks), the initial stock of spare parts, and the initial munition requirements. It also refers to the equipment of the unit's supporting infrastructure. Equipment generates both investment costs (for purchase) and operating and support costs (discussed under peacetime activity below). Force structure changes can involve the procurement[11], retirement, modification, and transportation of equipment. Although requirements are well defined in service documents, net equipment costs associated with force structure change are not always obvious. For example, one alternative in the AH-1 to AH-64 modernization caused a higher net increase in the AH-64 inventory (because the equipment was coming off of production) than another alternative (which obtained some of the new equipment from deactivating units). Further, support equipment can sometimes be shared or taken from inventory (reducing new purchase requirements). On the other hand, force structure changes can sometimes generate modification costs. For example, a recent proposal to transfer KC-135s from the Active Air Force to the Reserve forces required placing new engines in the aircraft to match maintenance capabilities in the reserves.

13. How do peacetime activities change?

The peacetime activities of a unit include both the functions performed exclusively for the benefit of the unit itself (e.g., unit training) and those external functions that serve, at least in part, larger tasks or other parts of the force (e.g., cargo aircraft provide airlift services). Peacetime activity, in most cases measured by equipment operating tempo, generates costs from the operation of equipment (e.g., consumption of fuel) and the employment of civilian personnel, and, for equipment intensive units (e.g. that use aircraft or ships) can often far exceed the cost of personnel. Operating tempo can often change due to the alteration of force structure. For example, transferring units from the active to reserve forces typically (but not always) means reducing operating tempo. However, the analyst should not assume that a change in peacetime activity is always the result of the force structure change; it can also

---

[11] Procurement costs of major unit equipment is typically not included because that decision rarely hinges on active/reserve force structure issues.

result from a separate decision to reduce the number of tasks for which units train. In addition, the calculation of *net* changes in activity (and cost) requires that the analyst consider whether apparent changes in the peacetime activity are compensated for by non-specified units. In the C-141 example, one alternative required increased flying in non-specified units to maintain airlift services. Finally, changes in peacetime activity can also signify changes in capability. For example, the reduced operation of FF-1052 frigates when transferred to the reserves eliminates their potential use as a forward presence in peacetime.

14. How does mission capability change?

Mission capability refers to the set of wartime activities that units are expected to conduct, and to the units' ability to perform those missions. Information about capability is important in the cost analysis of force structure change because decisionmakers are often presented with unequal alternatives (e.g., see the description of differences in the C-141 example above). Capturing differences in capability among unequal force structure alternatives provides a context for the results of the cost analysis and a way of associating an output with a price tag. This, in turn, facilitates the integration of the cost results with those concerning effects on US defense posture.

Although directly measuring units' wartime mission capability is typically beyond the scope of cost studies, indicators of capability can capture significant information about output. Some of the relevant information (e.g., a description of the force structure change and the changes in resource and activity levels) is provided by answering other questions on this list. Other relevant indicators of capability change include changes in unit mission statements or deployment schedules, and changes in measures of unit performance, such as unit status ratings.

15. How will the change affect basing structure?

The basing structure refers to the land on which a unit is located and to the facilities it uses. It also refers to the land and facilities of the unit's supporting infrastructure. Force structure changes can involve the sale or purchase of land; the construction, rehabilitation, or mothballing of facilities; the opening or closing of entire bases; or the realignment of units on those bases. It can also involve the avoidance of the costs connected with any of those activities. Facilities include: those used in connection with unit equipment, such as hangars, runways, docks, maintenance buildings, supply facilities, and fuel storage facilities; those used to support the unit's personnel, such as dining halls, commissaries, and barracks; and those used for overall administration of the base. As with other resources, the costs of basing changes can vary widely because they depend not only on requirements, but also on what excess capacity already exists and on peculiar, base-specific costs. For example, in the C-5 case, the cost of locating a unit at a new base was over $100 million more than locating the same unit at an existing C-5 base with excess capacity.

An Additional Level of Detail

In addition to a list of questions and a discussion of examples, the guidelines might also include an additional level of detail, one that supplies support for analysts in *answering* the questions posed. We will consider one example here.

Question 2 on the list asks the analyst to add indirectly affected units to under-defined problem description. Below, we describe a decision-tree approach (or tool) that assists the analyst in answering that question. The decision-tree approach aids in making the inquiry *systematic*, the key to the analysis of indirectly affected units.

Answering the question requires following the resource and mission trails of directly affected units. The procedure developed calls for inquiring into the *source* of resources, activities, and missions in force structure changes when directly affected units are gaining them; and into the *disposition* of resources, activities, and missions in force structure change when directly affected units are losing them. If any resource, mission, or activity has its source or disposition in another unit or part of the force, the analyst must ask whether that other unit properly belongs within the scope of the problem at hand. The answer will depend on whether the changes in the other units can be considered predetermined and their costs sunk, or whether the changes are a logical cause or consequence of the force structure change under consideration.

The clearest example is provided by a unit's major equipment. If the major equipment from a unit deactivation were to be used to modernize the equipment of another unit, both units might well be properly included as part of the force structure change, even if the change was described as a simple deactivation, and the modernized unit was not identified in the problem definition. Further, if the old equipment of the modernized unit goes to yet other units (instead of being scrapped or placed in idle capacity), those other units might also be properly included in the problem definition.

The complete process of determining whether units should be added to the scope of the problem can be portrayed in the form of decision trees that distinguish the 5 basic types of unit change--changes in wartime mission, peacetime function, manning, equipment, and bases. Figure 1 considers the case of the subtraction of those items, as occurs when a unit is deactivated. Following the logic for each of the 5 categories for each unit originally named in a problem, the analyst can systematically determine whether additional units ought to be added to the problem description.

To see how the decision tree works, consider again the ripple effect caused by the modernization of a unit's major equipment (examples of equipment modernization are described in the C-141 and AH-64 cases considered above). If the unit's old weapon system is placed in inventory, disposed of, or sold, the resource trail would end with no other units involved. Following the decision tree along those options in Figure 1 leads to the short vertical line, indicating an end to the inquiry about major equipment with no new units added. However, if the displaced weapon system was used to modernize other units, with still older equipment, or if it was used to augment other units of the same type to a larger size, then the analyst needs to consider whether those units should also become a part of the problem definition, and their costs included in the analysis. This is indicated in Figure 1 by the "move" and "another unit" options, as well as by the shaded box that asks whether units should be added.

Following Figure 1 in the area of "peacetime function" would have led to the addition of the active squadrons in the C-141 case who took over the military airlift function of the deactivating unit. In addition, following some of the other paths would have assured the analyst that other units did not have to be added. For example, in the area of "basing", they

168

could have determined that none of the cases involved the closure of active bases.[12]   Further, without Figure 1, important indirect effects could easily be missed.   For example, in the AH-64 case discussed above, one might have tended to assume (erroneously) that, because that particular attack helicopter was fairly new, all those going to the Guard were coming from new production rather than from existing active units.

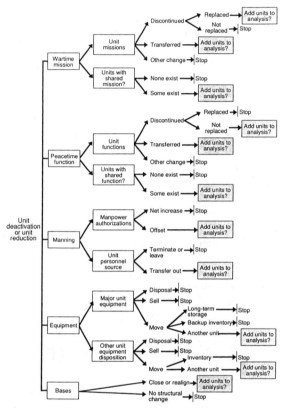

Fig. 1 —Tracing the Effect of Unit Deactivation or Reduction
on Problem Scope

CONCLUSIONS AND BROADER APPLICATIONS

The force structure costing guidelines have been designed as a tool for planning cost studies, a method for obtaining an overview of the force structure problem before becoming immersed in the details of cost calculations.   The guidelines address the three most common

---

[12] If the base on which a unit was located was logically part of the problem definition, the analyst may have to deal with support units in much the same way as combat units. For example, consider the example of closing Norton Air Force Base. In that case, the analyst would find it necessary to add units called the Air Force Inspection and Safety Center, the Audit Agency, and the Audio Visual Service Center--missions that were located on the base alongside the combat squadrons. The analyst would then have to ask about the disposition of those missions and about their manpower, equipment, and activity level.

pitfalls of the cost analysis of force structure change: omitting *indirectly* affected parts of the force from the analysis, ignoring or under emphasizing critical elements of the cost of (or savings from) *transition*, and incorrectly calculating the *net* effects of a change. Use of the guidelines will lead to more accuracy in force structure cost estimates, not necessarily because alternative estimates of generally-stated problems will converge, but because what specific problem is addressed by each estimate and what the results mean will become clear. For the inexperienced analyst, the guidelines can serve as a comprehensive guide; for the expert, they will more likely serve as a checklist to ensure completeness of work. In either case, analysts will be better equipped to plan a detailed analysis of the cost of force structure change, or to properly qualify the results of a quick estimate.

In addition to the primary benefits, the guidelines are likely to have beneficial side effects that improve force structure decision making. First, the guidelines developed in this article will work in conjunction with other improvements in force structure costing, such as improvements in the cost methodology itself.[13] Even quite sophisticated models of force structure change will require that the analyst complete a problem definition exercise before using the model. Second, following the guidelines will facilitate the identification of valuable new alternatives. For the decisionmaker, the process of carefully and explicitly specifying a decision will naturally lead to ideas about alternative methods of implementation. For the cost analyst, unresolved ambiguity about how a change will be implemented will naturally lead to the creation of cases and the testing of tradeoffs that explore the consequences of alternative resolutions of that ambiguity.

Third, the procedure will facilitate the presentation of cost results, because answers to those same questions addressed at the beginning of a study are likely to provide appropriate information for placing the results in the proper context. Thus, in addition to the bottom-line cost of a force structure change, results should also include information about changes in resources, activities, and missions within the force structure, information that the question list provides.

The *particulars* of the guidelines presented in this paper will be of interest primarily to those concerned with estimating the cost of altering military force structure. However, the *approach* used to create the guidelines goes beyond the single issue of military force structure. At the most fundamental level, the guidelines represent a strategy for assisting analysts in navigating through the problem formulation stage of any cost analysis. The guidelines work to define the problem scope, to identify major issues and critical assumptions and important constraints, and to fill out and (if necessary) expand the nominal alternatives presented to the analyst.

As a problem formulation tool for cost analyses, the guidelines have several identifying characteristics. First, rather than a general guide to setting up cost analyses[14], the approach calls for issue-specific guidelines; that is, aids that incorporate the main analytic issues that

---

[13]Improvements in force structure costing methodology, sponsored by both the Services and by OSD, are currently underway. For example, RAND is currently completing a two year project to develop and implement an automated system for costing changes in force and support structure in the Department of Defense. This project is being conducted for the Office of the Assistant Secretary of Defense, Program Analysis and Evaluation.

[14] For an example of such a guide, see Gary Massey, *Cost Considerations,* published as Chapter 10 of Hugh Miser and Edward Quade, eds., "Handbook of Systems Analysis: Craft Issues and Procedural Choices," Elsvier Science Publishing Co., Inc., New York, 1988.

arise in a particular subject area. Second, the guidelines approach replicates how *expert* analysts go about developing a vaguely worded proposal into one suitable for analysis; namely by asking a series of focused questions. Third, the approach uses well-structured examples under generalized guidelines to supply the required specificity and to teach users how to apply the tool in the context of their own work. Finally, the approach allows for supporting methodologies and data bases to support the analyst in answering the questions in the context of new situations.

While every costing effort begins with a process of problem formulation, in most cases the process amounts to a mental exercise undertaken by the analysis involved. When is the effort warranted to construct a formal set of guidelines? Clearly, one requires a setting in which a given type of problem is expected to arise on a continual basis in a relatively well-defined context. One-shot analyses have no need to produce generalized tools. Further, there must be some evidence that current methods of formulating the problem falter in defining critical problem parameters. The reasons for the inadequate problem definitions could vary widely-- for example, it might be due to widely varying (and perhaps inadequate) methodologies, use of untrained analysts in completing studies, time pressures or other restrictions in completing a study, or an advocacy environment which contains incentives for hiding critical value-laden assumptions--but the results will always be a large variance in results when more than one analyst independently addresses the "same" problem. It appears to this author that the scenario described above will fit a large number of real-world situations.

APPENDIX--GUIDELINES SUMMARY

For easy reference, the full question list appears below, organized into three subject areas--the change in force structure, changes during the transition period, and net changes in DoD resources, activities, and missions. The questions apply to each alternative considered in a force structure decision.

THE CHANGE IN FORCE STRUCTURE

The first group of questions address the nature of the change and the units affected.

1. What are the characteristics of the "base case"?
2. What military units are affected, directly or indirectly, by this alternative?

    a. What are the named combat units involved in this force structure change?

    b. What are the combat or deployable support units that are indirectly affected through a shared mission?

    c. What units are affected indirectly through the transfer of resources, activities, or missions?

3. What is the type of each unit affected?

    a. What is each unit's component (active, reserve, guard)?

    b. What is each unit's description (MDS or more specific)?

4. What type of change will the affected units undergo?

    a. How many activations of new units?

    b. How many deactivations of existing units?

    c. How many existing units will undergo resource, activity, or mission changes?

5. Does the number of units increase, decrease, or stay the same?
6. Does the change affect the unit support structure?

    a. Are there any changes to personnel establishments or personnel programs (e.g., acquisition, training, bonus levels)?

    b. Are there any changes to maintenance establishments (e.g.,depots, intermediate maintenance) or other central logistics organizations?

    c. Are there any changes to basing establishments (e.g., base openings, closings, realignments)?

    d. Are there any changes to headquarters or administrative organizations?

7. Are there other changes, unrelated to force structure, that are included in the problem definition?

CHANGES DURING THE TRANSITION PERIOD

The next set of questions addresses the nonrecurring transition costs associated with the proposed change.

8. What are the transition tasks and the associated nonrecurring costs or savings?

    a. What are the administrative and planning tasks, and what will they cost?

    b. What are the personnel processing tasks, and what will they cost?

    c. What are the equipment processing tasks, and what will they cost?

    d. What are the tasks associated with changes in facilities, and how much will they cost?

9. To what extent can transition costs be saved by making use of untapped resources and excess capacity?

a. What proportion of new personnel will have prior-service experience?

b. What proportion of support equipment and spares can be shared with an existing unit?

c. To what extent can the unit draw on excess capacity or more efficient basing configurations to reduce a unit's facilities requirements?

10. What role does timing play in costs?

a. How much time will it take for each unit affected to transition to the new force structure?

b. Within the transition period, what are the year-by-year costs?

c. When transitioning from old to new force structures, is there an overlap in the operation of units that are expanding with those that are contracting?

NET CHANGE IN DOD RESOURCES, ACTIVITIES, AND MISSIONS?

The third set of questions is intended to identify net changes in resource and activity levels that drive cost. In addition, indications of changes in military capability are addressed.

11. How does personnel endstrength change?

a. How many personnel positions are affected, including those not defined in unit manning documents, but nonetheless connected with the unit mission?

b. Will total endstrength increase, decrease or stay the same?

c. Will the personnel composition (e.g., grade   distribution) of the force change in ways that will significantly affect cost?

12. How does equipment change?

a. Does the number of major end-items of equipment increase, decrease, or stay the same?

b. Does the type of major equipment change?

c. What are the net changes in requirements for mission-specific equipment and munitions (e.g.,missiles, WRM, spares, support equipment)?

d. What costs for mission-specific equipment and munitions can be avoided due to the force structure change?

e. What modifications are required to major end-items of equipment as a result of the force structure change?

13. How do peacetime activities change?

a. Does the level of peacetime activity (in most cases measured by OPTEMPO) increase, decrease, or stay the same?

b. Does the distribution of peacetime activity across weapon systems and other types of equipment change in ways that significantly affect cost?

14. How does mission capability change?

a. Does the force structure expand, contract, undergo modernization, shift toward one component or the other, or otherwise change?

b. Is there a change in unit resourcing or maintenance practices?

c. Do unit mission statements change?

d. Is there a change in readiness ratings, deployment schedules, crew ratios or other descriptors of unit output or performance?

15. How will the change affect basing structure?

a. Will facilities or land at the base increase, decrease, or otherwise change?

b. What facilities costs are avoided due to the force structure change?

c. Are bases opened, closed, or realigned as a result of the force structure change?

REFERENCES

Air Force Magazine, May 1990.

Department of the Navy, *A Report to the Congress on the navy's Total Force,* February 1984, p.IV-4.

Gotz, Glenn A., et al., *Estimating the Costs of Changes in the Active/Reserve Balance*, R-3748-FMP/PAE/JCS, September 1990.

Kostiuk, P.F., *Cost Analysis of Selected Units in the Marine Corps Active and Reserve Components*, CRC 519, Center for Naval Analyses, January 1984.

Massey Gary, *Cost Considerations,* published as Chapter 10, Hugh Miser and Edward Quade, eds., "Handbook of Systems Analysis: Craft Issues and Procedural Choices," Elsvier Science Publishing Co., Inc., New York, 1988.

Schank, John F., et al., *Cost Analysis for Reserve Force Change*, R-3492-RA, 1987.

Schank, John F., et al., *Cost Element Handbook for Estimating Active and Reserve Costs*, R-3748/1-FMP/PAE/JCS, September, 1990.

Shanley, Michael G., *Active/Reserve Cost Methodology: Case Studies*, R-3748/2-PAE/FMP, March 1991.

Shanley, Michael G., *Guidelines for Planning the Cost Analysis of Active/Reserve Force Structure Change*, R-4061-FMP/PAE, 1992.

"The Part-Time Military," *National Journal*, March 4, 1989, page 519.

The Washington Post, editorial page, Sunday, April 6, 1989.

IV. Systems Cost Analysis

Cost Analysis of Prototyping Major Weapons Systems

Karen W. Tyson, J. Richard Nelson, D. Calvin Gogerty,
Bruce R. Harmon and Alec Salerno
Institute for Defense Analyses, 1801 North Beauregard Street,
Alexandria, VA 22311

INTRODUCTION

IDA performed a major study of acquisition for the Department of Defense, Effective Initiatives in Acquiring Major Systems. The study concerned the highly-publicized cost and schedule overruns that have plagued defense programs. Since the 1960s, the Defense Department and defense contractors have pioneered reviews and management initiatives to improve program outcomes. IDA reviewed program outcomes to determine whether there is a trend toward better outcomes overall, whether any specific management initiatives have improved outcomes, and what improvements could be made.

The study team found no broad improvements in aggregate outcomes. However, some of the initiatives IDA examined appeared to have potential for improving outcomes [1]. Prototyping was found to be effective, and IDA recommended that guidelines similar to those in use for multi-year procurement be established for prototyping.

The extent of prototyping is roughly counter-cyclical with the DoD budget. When the budget is ample, there is little prototyping, and when the budget is tight, there is more. For example, there was considerable prototyping in the periods of build-down after the Second World War and the Korean War. During the early 1960s, there was little prototyping, as the Kennedy administration believed that systems analyses could take the place of prototypes. Less than a third of major systems were prototyped. There was concern about paying for a prototype, finding problems, and then being left with no program or resources to fix the problem.

In the early 1970s, Deputy Secretary of Defense David A. Packard emphasized the importance of prototyping in a fly-before-buy strategy. Around half of major systems were prototyped [1]. During the early 1980s, when the Reagan buildup occurred, once again the defense budget increased relative to GNP, and there was less prototyping. The Packard Commission report in 1986 again called for more prototyping.

Consideration of prototyping is especially timely now for a number of reasons:

1. A decreasing real defense budget increases pressure on weapon system developers to make their programs more predictable and financially viable. Prototyping can provide these benefits.

2. As a consequence of lower overall budgets, reduced funding is available for major acquisition programs. Prototyping two systems is often cheaper than buying one.

3. Fewer new starts are anticipated in this lower defense budget climate. Therefore, the programs that are started may tend to be "all eggs in one basket" projects. The few new programs that are funded are likely to carry a great deal of technical risk and to push the state of the art. Since chances to win a

bid are becoming increasingly rare, there is a great deal of pressure to underestimate cost and schedule. Prototyping encourages realism in technology, cost, and schedule.

4. The ability of the government and contractors to sustain the defense technology base is in question. If not enough work is forthcoming from the DoD, then manufacturers will leave the industry. More importantly, new ideas will not be forthcoming from the technology base, and design teams will wither away. Ben Rich, head of Lockheed's Advanced Development Projects said, "Kelly Johnson [his predecessor] developed 47 different airplanes in his 50 years. In my 40 years, I developed 27 different airplanes. My young engineer today is going to be lucky to see one project–an ATF." [2] Prototyping can help to keep design teams together.

5. Threats to national security are changing as a result of the changes in Eastern Europe, and they are much more difficult to predict.

6. Technical sophistication is increasing. More sophisticated equipment carries even higher technical risk and risks of cost and schedule growth. Integration is becoming more complicated. Software costs are becoming a major part of system costs, and software projects have been difficult in the past. Making the transition from design to production is also a major concern, particularly if early research and development on manufacturing technologies are not addressed by the defense industry.

WHY NOT PROTOTYPE EVERYTHING?

Detractors of prototyping suggest several reasons why prototyping should not be undertaken. These include:

1. Takes too long. Decisionmakers may view the prototype phase as added on to the schedule without a prototype, when in fact an advanced development prototype probably saves time in EMD. (Prototyping can save a great deal of EMD time if it solves technical problems early on.) Moreover, even if decisionmakers want to take time savings into account, they may lack tools or models that allow them to do so.

2. Costs too much. Analogously with schedule, detractors regard the up-front cost of prototyping as an obstacle.

3. Slows momentum of the program. By conducting a pre-EMD prototype, program managers and others argue that necessary technical momentum is lost, and getting to initial operational capability (IOC) will take additional time.

4. Delays funding commitment. Along with momentum, major funding commitments tend to be delayed while prototypes are built and tested. In the present acquisition culture, large funding commitments (as in EMD) are viewed as necessary to "lock-in" support. Nevertheless, there is an alternative view that suggests that delaying large funding commitments allows the government to keep its options open.

5. Quantitative benefit not documented. The evidence on prototyping from the literature consists mainly of case studies and qualitative observations.

DEFINITION

We define a prototype as hardware used for testing, that is built before engineering and manufacturing development (EMD), previously called full-scale development (FSD). A prototype may have full or partial capabilities and may be full or partial scale. A prototype is used foremost to reduce technical risks, and then to reduce risks with respect to cost, schedule, or operational suitability. Decisionmakers gain information about the feasibility of a concept, the feasibility and cost of a design, and the feasibility, cost, and operational suitability of a particular design.

If an acquisition program is to be successful, potential design problems need to be identified and resolved as early as possible. Such problems can affect the performance and technical characteristics of the weapon, its development schedule, and its development, production, and support costs. A first step is to identify the sensitivity of the weapon's performance and technical characteristics, development schedule, and costs to perturbations in the weapon's design. For those perturbations that result in significant changes in the weapon's performance or technical characteristics, development schedules, or costs, possible risks should be identified and resolved through the construction and testing of prototypes. The earlier that this can be done in the development process, the less will be the subsequent required revisions to the design, manufacturing processes, and already completed parts of the system.

The primary purpose of prototyping is to reduce technical risk. Prototypes can be used to answer three technical questions. The three questions, which are not mutually exclusive, are:

o   Is the concept feasible?–Proof of Concept

o   Does the design work the way it is supposed to work?–Proof of Design

o   Does the system provide a militarily useful capability?–Proof of Mission Suitability

We examined prototypes for major weapon systems, corresponding to DoD budget categories 6.2, 6.3A, and 6.3B. The analysis focused on 6.3B prototypes. Our definition does not include 6.4 EMD test articles.

PROTOTYPING EXAMPLES--THE HARRIER AND THE ATF

The program that eventually resulted in the AV-8B Harrier aircraft benefited from all three categories of prototyping. The British P-1127 program in 1960 demonstrated the concept and the technology needed and was low-cost.

Proof of design was tested by the XV-6A in 1965, a program that led to the Marine Corps AV-8A in 1971. This was also a relatively low-cost program.

The AV-8A had limited mission capability and an improved version was required by the Marine Corps. In 1976, the YAV-8B prototype underwent mission demonstration testing that demonstrated a doubling of the payload-range capability of the AV-8A. This led to the AV-8B in 1980. The YAV-8B prototype was a high-cost program relative to the earlier prototypes.

A recent example of the benefits of prototyping in an actual program is the winning model for the Advanced Tactical Fighter, the Lockheed YF-22. The ATF is taking major technical leaps in multiple areas. The competitive prototype phase allowed exploration and resolution of many technical issues while spending was relatively low. In several key areas--the use of composite thermoplastic/thermoset materials, aerodynamics/configuration blending for low observables, and software, to name a few-- prototyping changed the way designers thought. The general manager for the program said that the prototype phase yielded information that probably prevented a pretty big cost and schedule problem.

QUALITATIVE AND QUANTITATIVE INFORMATION

Acquisition managers get two types of information from prototyping–qualitative and quantitative.

The qualitative information can include both design and program information. For example, in the prototype phase for the Lightweight Fighter (later the F-16), the fly-by-wire control and autostabilization system was refined and proven to work [3]. Questions such as: Can a missile achieve lock-on? Can a VTOL aircraft hover in controlled flight? What is the ground effect of vertical engines? Does the guidance system work? can be answered relatively inexpensively. In addition, prototyping yields programmatic information, such as whether contractor teams mesh well and, if there are competing teams, which group has the best design approach.

Quantitative information from prototyping includes performance, schedule, and cost dimensions. Required performance characteristics can be validated through the testing of a prototype, or the requirements can be changed to fit what can reasonably be achieved. Acquisition managers can also learn how long a program will take and how much it is likely to cost.

We cannot evaluate all the benefits of prototyping in a quantitative fashion. The qualitative benefits of prototyping are by definition not quantitatively measurable. In addition, one of the quantitative benefits, performance, is multi-dimensional and has different dimensions across equipment types. However, schedule and cost are measurable, and by measuring planned vs. actual schedules and costs, we can compare program outcomes across equipment types.

DATA USED--WHAT AND WHY

We were fortunate to have data on a large sample of major acquisition programs–those meeting the dollar threshold for filing Selected Acquisition Reports (SARs)–from a number of past IDA studies. The data we had were particularly appropriate for a study of prototyping, since they have uniform outcome measures for a wide variety of programs [1, 4, 5, 6]. For evaluation of cost and schedule growth outcomes, we used data on the 52 programs that had at least three years of data and that were not canceled, e.g., programs that bought or will buy at least three-quarters of the number of items planned. The tactical aircraft cost-estimating relationships (CERs) are from the IDA tactical aircraft development cost study [7], and the munitions CER is from Yates, Waller, and Vaughn [8]. There are some limitations in the data. We could not fully identify subsystem prototypes. Because we did not have cost growth measures for them, we omitted ship and vehicle programs.

OUR YARDSTICK

Prototyping helps primarily to reduce technical risks. However, such an impact is difficult to measure because the technical characteristics are unique to each equipment type and program. If a program proceeds well from a technical standpoint, then it is much less likely to encounter schedule and cost problems. Cost and schedule problems are measurable.

Because weapon systems are very dissimilar, analysts search for a common yardstick to measure program success. Over the last forty years, much has changed in the weapon acquisition process. However, there are common threads.

Since the late 1960s, current estimates of program cost and schedule for major programs have been reported in Selected Acquisition Reports. These current estimates are compared with original estimates at Milestone II (the EMD decision meeting) to determine how much cost and schedule growth the program has had. Programs that had a high level of cost growth are judged to be less successful than programs that had less cost growth. Analogously, programs that took much more time than planned from Milestone II full-scale development (FSD) (now EMD) to initial operational capability (IOC), are judged to be less successful, while those that met their schedule targets are thought of as being more successful. Such measures have the virtue of being index numbers that can be used to compare diverse systems using a similar benchmark. We adjusted cost growth for changes in production quantity and inflation.

THE IMPACT OF PROTOTYPING: THE EVIDENCE

1. Cost

a. Development Cost Growth Lower. Development cost growth was significantly less for prototyped programs than for non-prototyped programs. Thus, DEM/VAL prototyping allows program managers to make a more educated cost estimate at the time of Milestone II. Average cost growth was 62 percent for non-prototyped systems, and only 17 percent for prototyped systems. The effect was smaller, 29 percent vs. 17 percent (and not statistically significant), for all aircraft programs. (Among the tactical aircraft programs, non-prototyped programs grew by 18 percent and prototyped programs by 12 percent.) The tactical munitions showed the greatest payoff to prototyping. For non-prototyped programs, development cost more than *doubled* from its plan at EMD. Prototyped munitions had only 21 percent development cost growth.

We also tested whether program size had a confounding effect on development cost growth, independent of prototyping. Larger programs tend to have lower cost growth generally, perhaps because of increased management attention. In the aggregate and in tactical munitions, program size did have a significant negative effect on development cost growth. Nevertheless, prototyping remained as a significant factor in reducing development cost growth, independent of program size.

b. Fewer Unplanned EMD Articles. Prototyped programs have significantly fewer unplanned EMD articles. On the basis of prototype testing, program managers are able to make a better estimate of how many EMD articles they will need.

c. Production Cost Growth Lower. While the difference was not statistically significant, production cost growth was less for the prototyped systems than for the non-prototyped. Average cost growth was 55 percent for non-prototyped systems, and only 29 percent for prototyped systems.

d. Levels of Development and Production Cost. To examine the effect of prototyping on the levels of development and production costs, we turned to a standard tool of cost analysis, cost-estimating relationships (CERs). CERs relate technical characteristics of a weapon system to its development or production cost. We examined the residuals of the CERs to determine whether there was any significant difference between prototyped and non-prototyped systems. If we found that prototyped systems had significantly higher residuals, this would indicate that a system with given technical characteristics would cost more if it were prototyped. Conversely, if we found that prototyped systems had significantly lower residuals, it would indicate that prototyped systems generally cost less than non-prototyped systems.

We were able to perform the tests for three equations: a tactical aircraft airframe full-scale development CER, a tactical aircraft production CER, and a tactical munition full-scale development CER. For tactical aircraft airframes, there is no significant difference in either development or production costs that could be explained by prototyping. In the case of tactical munitions, there is no significant difference in development costs between prototyped and non-prototyped systems. (We were unable to locate a sufficiently aggregated CER to test munitions production costs). Thus, the available evidence on total costs suggests that prototyped systems of equivalent technical capability do not cost significantly more or less than non-prototyped systems.

2. Schedule

Overall, prototyped programs took 2 years longer than non-prototyped programs from Milestone I to IOC (significance level = .06), but prototyping made no difference in the time from Milestone II to IOC. For the aircraft, there was no statistically significant difference in either interval. Prototyped aircraft took slightly less than 9 months longer from Milestone I to IOC (117 vs. 108 months), but Milestone II to IOC times were virtually identical (69.6 vs. 70.4 months). The prototyped munitions took over 2 years longer than non-prototyped munitions (135 vs. 104 months), but the difference was not statistically significant. (Moreover, the more complicated munitions were prototyped. When we control for this relationship, the time difference decreases to 1 year.) The length from Milestone II to IOC was actually 5 months shorter for the prototyped munitions, but again not statistically significant.

Thus, prototyping may take some additional time. This time, however, must be weighed against the gains in cost and technical predictability. In addition, the extra time occurs at a time in the program when spending rates are low.

3. Diverse Strategies among Weapon Types

The two equipment groups in our study that had the most prototyping were aircraft and tactical munitions. We observed very different strategies regarding the prototyping of these two groups. Among the aircraft, the systems pushing the state of the art the least (such as the F-5E and the F-16) were prototyped, while others that were more technically difficult (like the F-14) were not. In the munitions, the opposite occurred. Systems with a high level of technical "reach" like Hellfire, HARM, and Harpoon were prototyped.

The strategy used for munitions was the more successful of the two. Munitions are often high-risk programs in general. They are less glamorous than aircraft and therefore seem to get less management attention. Perhaps the building and testing of a prototype serves to focus attention on the program. In any event, the munitions strategy was strongly successful. We would expect the munitions with high technical reach to have higher cost growth than those with low reach. In fact, those complicated munitions that were prototyped did *better* than the simple ones that were not prototyped.

In the aircraft, by contrast, the prototyping strategy did not seem to be as successful. However, when we remove the helicopters, which had generally higher cost growth regardless of prototyping strategy, the remaining prototyped aircraft have significantly lower development cost growth than the non-prototyped aircraft.

RECOMMENDATIONS

Prototyping enhances the credibility of major programs, particularly given the tendency to underestimate technical risk. In general, the earlier the prototyping strategy is undertaken, the better. For all-new systems, the concept demonstration phase is particularly important. Operational suitability prototyping is particularly important in times of budget crunch, since we are particularly eager to know whether a system will work significantly better than what we already have.

The type and extent of prototyping to be done also depends on the nature and extent of risk in the program. If the risk is largely technical, then concept and design prototyping are the most important. If the risk is that requirements are uncertain, then proving the technology is operationally suitable is most important. If there are concerns about production costs and producibility, it may be necessary to add a test of operational suitability with production article(s).

Prototyping is a leveraged investment. Spend small amounts of money now to avoid large surprises later. As rules of thumb for when prototyping makes sense relative to its likely payoff, we suggest that the prototype cost should be less than 25 percent of the EMD cost estimate, 10 percent of the acquisition cost estimate (EMD and Procurement), or 5 percent of the life cycle cost estimate, whichever is highest.

These rules of thumb can be adapted for technical risk and schedule criticality. If technical risk is high, then the cost estimates upon which these rules of thumb are based have considerable risk attached to them. For example, if technical risk is low, schedule is critical, and a prototype would cost 20 percent of EMD cost, then it would not make sense to undertake one. On the other hand, if technical risk is known to be very high, schedule is not critical, and a prototype would cost 30 percent of the EMD cost estimate, then prototyping makes sense.

NEED FOR FURTHER RESEARCH

Our quantitative analysis was not extensive enough to support development of a cost/benefit model for prototyping. Nevertheless, we believe that we have taken some important first steps toward such a model. A key element of such a model is a better measure of technical risk early in the acquisition process.

Our analysis suggests that it would be useful for cost analysts to capture the costs of prototypes to refine EMD and procurement cost estimates. The literature on this subject is surprisingly sparse.

It would also be useful to study the impact of prototyping in combination with other initiatives such as design-to-cost and contract incentives. In addition, the impact of a generalized strategy of prototyping across programs should be assessed. This should include its effect on competition and on the ability of industry to develop and produce new, technologically sophisticated weapon systems.

SUMMARY

The lessons learned from prototyping in the acquisition of major defense systems are overwhelmingly positive. Prototyping helps developers and users to understand the technical risks and uncertainty of the requirements.

The quantitative evidence about the benefits of prototyping is also strongly positive. Prototyping helps to reduce development cost growth, thereby offering improved program control. That effect is particularly pronounced for technically challenging programs. Development quantity growth, the need to build unplanned EMD articles, is significantly less for tactical munitions programs. The benefits of prototyping also carry over into production. Production cost growth is generally less for prototyped systems. These benefits come with some increase in development time. However, this additional time is not necessarily very long (and not statistically significant) for aircraft, and, for the tactical munitions, it may be more related to technical challenge than to prototyping. Prototyping is a leveraged investment. You are buying information relatively cheaply, early in the program, rather than discovering problems in EMD or production, when costs (and rates of expenditure) are higher. Our evidence suggests that prototyped programs do not cost any more than non-prototyped programs.

Prior studies of prototyping have been qualitative and have emphasized the uniqueness of each acquisition. Despite this uniqueness, policymakers should use consistent, clear lessons from past programs to set strategy for new programs. The quantitative and qualitative evidence we examined is clear. The payoff to prototyping challenging systems is large.

REFERENCES

[1] Tyson, Karen W., J. Richard Nelson, Neang I. Om, Paul R. Palmer, Jr., "Acquiring Major Systems: Cost and Schedule Trends and Acquisition Initiative Effectiveness." Institute for Defense Analyses, Paper P-2201, March 1989.

[2] Wartzman, Rick, "Designer of Stealth Fighter Says U.S. Runs Risk of Losing Technological Edge," *Wall Street Journal*, February 4, 1991.

[3] Smith, G. K., A. A. Barbour, T. L. McNaugher, M. Rich, and W. L. Stanley. "The Use of Prototypes in Weapon System Development." The RAND Corporation, R-2345-AF, March 1981.

[4] Gogerty, David C., J. Richard Nelson, Bruce M. Miller, and Paul R. Palmer, Jr., "Acquisition of Contemporary Tactical Munitions," 2 Volumes, Institute for Defense Analyses, Paper P-2173, October 1989.

[5] Harmon. Bruce R., and Lisa M. Ward. "Methods for Assessing Acquisition Schedules of Air-Launched Missiles," Institute for Defense Analyses, Paper P-2274, November 1989.

[6] Harmon, Bruce R., Lisa M. Ward, and Paul R. Palmer. "Assessing Acquisition Schedules for Tactical Aircraft," Institute for Defense Analyses, Paper P-2105, February 1989.

[7] Harmon, Bruce R., J. Richard Nelson, Mitchell S. Robinson, Katherine L. Wilson, and Steven R. Shyman. "Military Tactical Aircraft Development Costs." 5 Volumes, Institute for Defense Analyses, Report R-339, September 1988.

[8] Yates, Edward H., W. Eugene Waller, and Lem G. Vaughn. "A Parametric Approach to Estimating Cost of Development Engineering," Applied Research Inc., ARI/87 TM-387, September 1987.

Cost Estimating Relations for Space-Based Battle Management, Command, Control, and Communications Systems

Daniel A. Nussbaum
Naval Center for Cost Analysis

Elliot Feldman and Mark McLaughlin
MATHTECH

Everett Ayers
ARINC RESEARCH

Donald Strope
COST, Inc.

ABSTRACT

This report documents an effort to develop a database and cost estimating relation (CER), for Battle Management, Command, Control and Communications (BM/C^3) architectures.

Data on twenty-eight BM/C^3 communications systems were collected, and a common Cost Work Breakdown Structure (WBS) was devised to eliminate cost accounting disparities among these architectures and organize the data into a single database.

The data were subjected to statistical regression analysis in order to estimate a linear equation that relates the system integration costs to system technical and performance characteristics, such as weight, data flow speed, and number of nodes.

Much of the data collection and analysis to date has concentrated on a relatively homogeneous collection of space-based architectures, resulting in a CER that is particularly appropriate to such systems but not to the full gamut of BM/C^3 systems required for comprehensive cost estimation. Consequently, a recommendation arising from this study is to obtain data on additional, more heterogeneous, BM/C^3 architectures in order to expand the usefulness of the cost estimating tool under development.

INTRODUCTION

The purpose of this study is to develop a data base and cost estimating relations (CERs) for Battle Management, Command, Control and Communications (BM/C^3) architectures.

Effective C^3 requires significant resources for advanced sensors, displays, communication equipment, and computer processing equipment. To compensate for increased cost associated with the improved operational capabilities obtained from C^3 systems, and to provide greater interoperability, individual systems are frequently grouped and connected in network architectures that share information processing and transmission capabilities.

The approach in this effort is:

o   develop a Cost Work Breakdown Structure (WBS).
o   collect cost data and technical and performance characteristics on twenty-eight systems.
o   crosswalk the cost data collected for each system to the elements in the WBS.
o   subjected the data base to statistical analyses in order to develop CERs.

The paper first addresses the data base development effort including the data collection process and system descriptions. Next the paper documents the cost models that were researched and describes their applicability to BM/C^3 cost estimating tools. Then the paper describes the data analysis efforts and provides the foundation for the regression and other statistical analyses performed. Finally, conclusions and recommendation for future development of BM/C^3 cost estimating tools are presented.

DATA BASE DEVELOPMENT

We collected cost and technical data for twenty-eight systems. Thirteen of these systems are space based systems and were obtained from the NASA Goddard Space Flight Center (GSFC). Other organizations that contributed data were Naval Center for Cost Analysis (NCA), Defense Communications Agency (DCA), Strategic Defense Initiative Office (SDIO), and Space and Naval Warfare Systems Command (SPAWAR).

A Cost Work Breakdown Structure (WBS) provides a detailed plan for organizing the data collected. WBS's often display costs at a detailed level, with hardware broken down into specific systems and subsystems. This effort uses a coarser WBS in order to facilitate high-level cost trade-off analyses among alternative BM/C^3 architectures. Table 1 displays the Cost WBS used in this effort, listed by life cycle phases (R & D, Production and System Activation, and Operations and Support).

Table 1. BM/C^3 Cost Work Breakdown Structure

0.0   TOTAL SYSTEM

1.0   RESEARCH AND DEVELOPMENT
    1.1   PROGRAM MANAGEMENT

```
1.2  SYSTEM DESIGN & ENGINEERING
1.3  HARDWARE DEVELOPMENT & MODIFICATIONS
1.4  SOFTWARE DEVELOPMENT
1.5  SYSTEM INTEGRATION & TEST
1.6  INTEGRATED LOGISTICS SUPPORT PLANNING
1.7  SYSTEM DEMONSTRATION

2.0  PRODUCTION & SYSTEM ACTIVATION
     2.1  PROGRAM MANAGEMENT
     2.2  SYSTEM ENGINEERING
     2.3  HARDWARE PRODUCTION
     2.4  SOFTWARE PRODUCTION
     2.5  SYSTEM INSTALLATION & TEST
     2.6  INTEGRATED LOGISTICS SUPPORT
     2.7  IMPLEMENTATION & INITIAL TRAINING

3.0  OPERATIONS & SUPPORT
     3.1  PROGRAM MANAGEMENT
     3.2  SUSTAINING ENGINEERING
     3.3  HARDWARE MAINTENANCE, REPAIR, & SPARES
     3.4  SOFTWARE MAINTENANCE
     3.5  INTEGRATED LOGISTICS SUPPORT,
     3.6  PERSONNEL, & REPLACEMENT TRAINING
```

Within each of the R&D and Production/System Activation Life Cycle Phases, there were two groups of costs. The first was called "wraparound" cost, or the cost to integrate and test the components of a complete system; and the second was termed "hardware/software" cost. Wraparound cost was the sum of five elements while hardware/software cost was the sum of two elements, as follows:

```
┌─────────────────┐
│ Wraparound Cost │
├─────────────────┴────────────────────────────┐
│ 1.1 Program management                         │
│ 1.2 System Design and Engineering              │
│ 1.5 System Integration and Test                │
│ 1.6 Integrated Logistics Support Planning      │
│ 1.7 System Demonstration                       │
└────────────────────────────────────────────────┘

┌──────────────────────┐
│ Hardware/Software Cost│
├──────────────────────┴─────────────────────────┐
│ 1.3 Hardware Development & Modifications        │
│ 1.4 Software Development                         │
└─────────────────────────────────────────────────┘
```

A problem that often occurs in the data collection phases of cost estimating is that the cost WBS structure can be significantly different from the categories and cost breakdown items of data that are actually collected. In fact, none of the systems we obtained organized data in the same format as the BM/C^3 cost WBS. Consequently, after collecting the cost data for a given system, a "crosswalk" process was performed that mapped the cost breakdown items that were collected to the cost WBS items defined for this study.

Table 2 shows how some of these activities are associated with the WBS categories.

**TABLE 2. Cost Categories and Activities
Associated with Cost WBS Categories**

```
PROGRAM MANAGEMENT
            PROGRAMMING AND BUDGETING
            PLANNING
            ACQUISITION DOCUMENTATION
            WORK BREAKDOWN STRUCTURE
            COST ESTIMATES
            REVIEWS AND PROGRESS REPORTS
            TRAVEL
SYSTEM DESIGN & ENGINEERING/SYS. ENG./SUSTAINING ENG.
            TECHNICAL REVIEWS
            INTERFACE MANAGEMENT
            SYSTEM DEFINITION
            PRELIMINARY SPECIFICATIONS
            DEVELOPMENT ASSESSMENTS
            ENGINEERING TRADES
            CONFIGURATION CONTROL
            PRODUCTION ENGINEERING
            PRODUCT ASSURANCE
            INDEPENDENT VERIFICATION & VALIDATION
HARDWARE DEVELOPMENT & MODS./PRODUCTION/MAINT. & REPAIR
            PROJECT MANAGEMENT
            SYSTEM ENGINEERING
            PRODUCT ASSURANCE
            SUPPORT EQUIPMENT
            ASSEMBLY & INTEGRATION
            FIELD SUPPORT
SOFTWARE DEVELOPMENT/PRODUCTION/MAINTENANCE
            MANAGEMENT
            DEVELOPMENT
            INTEGRATION & TEST
            FIELD SUPPORT
SYSTEM INTEGRATION/INSTALLATION & TEST
            SYSTEM TEST AND EVALUATION PLANS
            SYSTEM TEST PROCEDURES
            SYSTEM TEST HARDWARE/SOFTWARE
            SYSTEM ASSEMBLY & INTEGRATION
            INSTALLATION & CHECKOUT
            SYSTEMS OPERATIONAL VERIFICATION TESTING
INTEGRATED LOGISTICS SUPPORT
            LOGISTICS DEFINITION AND PLANS
            PERSONNEL
            INITIAL & REPLACEMENT TRAINING
            SUPPLY SUPPORT
SYSTEM DEMONSTRATION
            TESTBED DEVELOPMENT
            DEMONSTRATION PLAN
            DEMONSTRATION
```

One of the accomplishments of the study was to identify a relatively small set of technical cost drivers. Thirty-five technical parameters were initially identified as candidate cost-driving variables for BM/C^3 CERs. This list was narrowed to the set of five system level and three unit level parameters listed below:

            o System Level
                - Platform Location or Type
                - Number of Input Types
                - Number of Output Types
                - Number of Nodes and Interfaces
                - Data Rate

o Unit Level
    - Size
    - Weight
    - Length

These eight parameters were considered to be readily available during the collection process, and they were considered sufficient to characterize the technical complexity of a BM/C³ architecture. Furthermore, as a general rule, such parameters ought to be available during data collection because they are usually estimated early in system development.

Table 3 provides a summary of twenty-eight systems for which we were able to collect cost and technical data. In many cases, it was also possible to collect details of the system descriptions and system block diagrams.

Table 3. BM/C³ Data Base

AIR FORCE Communications System 1
Defense Switched Network (DSN)
Defense Communications Operations Support Systems (DCOSS)
Defense Systems Communication System (DSCS)
Extra High Frequency (EHF) Information Exchange System (IXS)
Extra High Frequency (EHF) Satellite Communications (SATCOM)
Jam Resistant Satellite Communications (JRSC)
Microwave Landing System (MLS)
MK XV Identification Friend or Foe (IFF)
Space Based Systems (13 NASA systems)
NASA Operational Communications (NASCOM)
NAVY Communications System 1
SAFEGUARD
Space Defense Initiative (SDI)
Tacintel II
Tracking and Data Relay Satellite System (TDRSS)

During the data collection process, we obtained technical descriptions and parameters for each system. We extracted this information from technical literature, such as technical overviews, system specifications, and system design documentation. In some cases, program offices were willing to provide data only if the system name and description were not released. Since our statistical analysis deals with numerical data values, we agreed to these restrictions. The affected cases include one Navy system, one Air Force system, and the NASA space based systems.

The following paragraphs provide a brief description of each system. The system descriptions are provided to illustrate sample BM/C³ architectures and their similarities to each other and other candidate architectures.

Defense Switched Network (DSN)

The Defense Switch Network (DSN) support effort for the Defense Communication Agency (DCA) includes DSN System Management Support, DSN Networking and System Engineering, and DSN Acquisition and Transition.

Defense Communications Operations Support System (DCOSS)

The purpose of the DCS System Control project is to provide an integrated and secure multi-hierarchical structure for comprehensive system control of the worldwide DCS. Such a control structure will enable DCA operational management personnel to exercise real-time monitoring and control of the DCS switched and data networks and transmission system to ensure cost effective and timely service to DCS users. It is also designed to improve the capabilities and survivability of the DCS during wartime or crises stress conditions.

Extra High Frequency (EHF) Satellite Communications (SATCOM)

EHF SATCOM interfaces shipboard tactical data communications subscribers to the AN/USC-38(V)3 Navy shipboard EHF SATCOM terminal and controls data exchange between subscribers over EHF SATCOM links. EHF SATCOM links include those provided by Milstar and the EHF Follow-On (UFO) satellites. The Data Communications Controller (DCC) also provides interfaces to existing OTCIXS and TADIXS subsystems.

Microwave Landing System (MLS)

The Military MLS Avionic (MMLSA) will provide integrated MLS and ILS capability in fighter and other high-performance aircraft in the same space currently occupied by an ILS receiver. The avionics must be interoperable with any configuration of military or civil ILS and MLS ground equipment. This requirement will be met by using the International Civil Aviation Organization (ICAO) standard signal formats for both ILS and MLS.

NASA Operational Communications (NASCOM)

The NASCOM system is a collection of individual communication networks, transmission media, relay stations, tributary stations, interfaces and terminal equipment(s) established and operated by NASCOM management that is capable of interconnection and inter-operation to form an integrally identifiable functioning entity.

Tracking and Data Relay Satellite System (TDRSS)

The Tracking and Data Relay Satellite System is a satellite tracking and communication system that operates at both S-band (2025-2300 mHz) and at Ku-Band (3700-5115 mHz). The Multiple Access Service (MA) operates in the S-band communications frequency and uses a fixed antenna on the TDRSS satellite to communicate with many spacecraft at the same time. This service is designed for low-rate, long-duration users. Precise location of the spacecraft can be determined, but is not required for use of the MA services. The amount of data (band-width) that can be relayed to the Earth is limited to 50 kilobits per second.

LOTUS 1-2-3 was selected as the standard software to standardize data base entry to a single software package that could provide both data base and regression analysis capabilities. All cost data are normalized to FY90 dollars. NCA inflation rate tables were used to adjust the data. On a system by system basis, it was decided to select an average inflation rate for a given cost WBS category based on the distribution of the cost data across all the systems for that element.

After we performed the crosswalk for each system, we entered the data into a spreadsheet, thereby providing a separate documented history of the crosswalk process for that system.

Wraparound costs are the sum of items 1.1, 1.2, 1.5, 1.6, and 1.7 in the BM/C^3 Cost WBS. Table 4 shows that RDT&E wraparound costs range from 15% to 82% of total RDT&E costs and average 40%.

TABLE 4. RDT&E Wraparound Costs by Percentage

Project Number	RDT&E Wraparound Costs (% of RDT&E Total Costs)
1	−
2	57%
3	81%
4	52%
5	54%
6	45%
7	82%
8	31%
9	−
10	−
11	55%
12	75%
13	−
14	59%
15	−
16	33%
17	22%
18	20%
19	15%
20	19%
21	17%
22	35%
23	35%
24	21%
25	21%
26	29%
27	22%
28	−

COST MODEL RESEARCH

As part of the BM/C^3 Cost Estimating Tool Development study, we investigated existing cost estimating tools and techniques to determine whether they were applicable to BM/C^3 systems. We discovered life-cycle cost models such as the Defense System Management College's Cost Analysis Strategic Assessment (CASA) model, the Air Force's LCC-2 model, ARINC's Automated Cost and Budget Estimating Network (ACBEN), and many others. But none of these models dealt adequately with estimating wraparound costs. We did find, however, three sets of guidelines used by DCA, NASA, and SDIO. These guidelines estimate the engineering and integration effort to cost between 5% and 20% of hardware costs are described in the following paragraphs.

DCA Cost and Planning Factors Manual (DCA CPFM)

In the Defense Communications Agency (DCA) <u>Cost and Planning Factors Manual</u>, DCA Circular 600-60-1 of March, 1983 the following guidelines are provided for the estimation of communication system integration costs.

o   Integration and assembly costs are currently estimated as ranging from 5% to 20% of the total prime mission and auxiliary equipment acquisition cost. For routine systems using standard equipment, use the 5% factor. For new systems using equipment developed by many different manufacturers and of unusual complexity, use the 20% factor.

o   The planner should generally use a factor within this range unless the uniqueness of the project or other information dictates the use of another, more accurate relationship or estimating procedure, such as man-years and material expenses.

NASA CDOS Cost Analysis

In the MITRE Report, MTR-89Q00001, <u>Customer Data and Operations System (CDOS) Cost Analysis</u>, the following Estimating Approach was used to estimate Engineering and Integration efforts for NASA systems.

Cost Estimating Relationships (CERs), as percentages of Prime Mission Equipment (PME), were used to estimate the PME Integration, System Engineering and Program Management (hardware portion), System Test and Evaluation, Site Activation, and the System Implementation (hardware portion) elements of the WBS. Table 5 shows the factors used in each separate estimate.

Table 5. NASA Cost Factors

WBS Element	Cost Estimating Relationship
PME Integration	10% of PME Software and Hardware
System Engineering/Program Management (hardware only)	20% of PME Hardware
System Test and Evaluation	15% of PME Software and Hardware

Table 5. (CONT)

Site Activation   10% of PME Hardware

System Implementation   55% of PME Hardware
(hardware portion only)

Table 6 displays some basic factors employed in the SDIO Cost Estimation Model which are also used in the DCA Cost and Planning Factors Manual (DCA CPFM).

Table 6. BM/C³ Cost Factors used by DCA and SDIO

Quantity	Factor	Reference
System Engineering	10% of Prototype Manufacturing	DCA CPFM
System Test & Evaluation	D&V - 5% Prototype Mfg. FSD - 10% Prototype Mfg. Inv - 7% Prototype Mfg.	DCA CPFM
Architecture & Engineering	8% Prototype Mfg.	DCA CPFM
Project Management	10% Prototype Mfg.	DCA CPFM

We conclude that these cost factors and guidelines are applicable to BM/C³ systems from the standpoint of establishing a range of percentages to estimate engineering and integration effort for communication systems. The range of 5% to 20% of hardware costs is a useful value for comparison with estimated results. Since engineering and integration are part, but not all, of what we call BM/C³ "wraparound" costs, it is reasonable that we found in Table 2-4 that RDT&E wraparound averages 40% of hardware and software costs for our sample projects. RDT&E wraparound includes program management and logistics support and system demonstration in addition to engineering and integration. Therefore, if engineering and integration effort totals 5% to 20%, and if the other wraparound efforts are similar, these cost factors provide reasonable guidelines.

COST ESTIMATING TOOLS DEVELOPMENT

The initial regression database consisted of a sample of thirteen space based systems obtained from the National Aeronautics and Space Administration. The cost elements of the system-specific work breakdown structures were mapped onto the generic WBS as described above.

It was theorized that space communication systems R&D system integration costs can be modeled as a function of system performance and technical characteristics such as number of terminal nodes, data communication (baud) rate, and number of platforms (air, ground, space, sea). The costs that are assumedly driven by increasing complexity represent expenditures to develop, test, and organize the many components of these systems. Thus any measure of system complexity would reflect the difficulty and, ultimately, the labor and capital cost of performing such tasks. Therefore, the implicit form of this cost model is

system integration cost = f(system complexity).

For each of the initial Thirteen systems, the following technical data elements were collected:

- weight (in pounds)
- volume (in cubic feet)
- frequency ranges (UHF, VHF, etc.)
- number of nodes
- number of instruments
- data rate (baud rate)

Two new technical attributes were derived from the data:

- density (weight/volume)
- number of frequency ranges

In the case of density, we divided weight by volume and in the case of the number of frequency ranges, we counted the number of separate bandwidths that each system was capable of using.

Because of the low sample size (13 data points), there was concern that the standard errors would be too high to allow the stepwise entry of important explanatory variables. In searching for a way to prevent this, similar WBS elements from two successive stages of the life-cycle were pooled. We took advantage of the fact that the WBS represents stages of the system life cycle and reasoned that the influence of system complexity is felt throughout the life cycle. Therefore, we made the following assumption:

The technical and performance characteristics influencing the cost of R&D and Production/System Activation come from the same set.

Based on this assumption, we combined the data from each phase into one regression data set. We have done analyses, not contained in this article, to demonstrate that the pooling of the data in this way yields coefficients that are insignificantly different from those obtained from separate regression on each phase.

By the inclusion of Production/System Activation costs, the sample size was doubled from 13 to 26 data points at the expense of adding one explanatory variable, a "dummy", that was set to a value of (1) if the observation was of production/System Activation (P/SA) and (0) if the observation was of R&D costs. Because P/SA costs are significantly lower than R&D costs over all WBS elements, we expected to see the coefficient of the dummy to be significant and negative. This expectation was correct.

After evaluating a large number of trial models generated by stepwise regression, a promising two-equation model was found. By "two-equation" model, we refer to a situation in which one of the explanatory variables of the first regression equation is predicted by the second equation.

We examined the best one, two, three, and four variable two-equation models. We stopped at the best two- and four-variable model due to the small sample size. The general form of the models is:

[1]   HS = f (NODES, SDUR, WGT, BANDS, i)

[2]   WRAP = g (HS, WGT, i   ), i = 0,1, where

NODES =   number of nodes and instruments

SDUR =   dummy for Short Life-cycle systems 5 & 6

WGT   =   satellite weight in pounds

BANDS =   number of frequency band

i   =   dummy to differentiate between R&D (i=0) and

PRODUCTION & SYSTEM ACTIVATION costs (i=1)

Table 7 briefly outlines one of the CERs we have estimated. The udependent variable in Table 7 is wraparound cost. It depends upon two explanatory variables: HS (the cost of Hardware/software R&D in FY90$K, and a technical parameter WGT (the weight of the satellite in pounds). The coefficients are the marginal effects of the independent variables, i.e., the rate of change in cost with respect to unit changes in the explanatory variables.

The numbers in parentheses just under each coefficient are standard errors, a measure of error for the coefficient. The ratio of the coefficient to its standard error is called a t-value. The t-value determines the statistical significance of the coefficient, i.e., whether its value is truly non-zero, or if its value is non-zero purely by chance.

The intercept term can be considered, in part, a fixed cost and, in part, the variation of any neglected explanatory variables.

Table 7: Interpretation of the CER

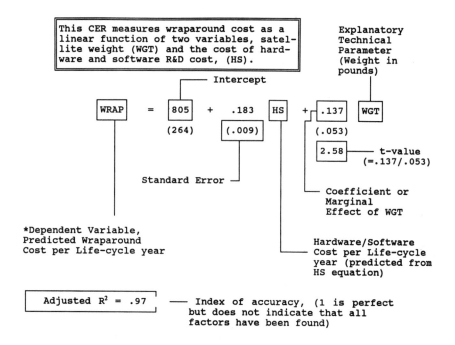

Because the dependent variable of equation [1] above (HS) lies on the right-hand-side of equation [2] above, the least-squares assumption that the residual vector is orthogonal to the vectors of the explanatory variables is violated, causing the Ordinary Least Squares (OLS) estimator to be biased. Table 8 presents and compares the Two- Step Least Squares (2SLS) as well as OLS estimates.

In Table 8, the coefficients of equations [1] and [2] are presented. Each row in the table represents a variable in either of the equations. The first two columns are the Ordinary Least Squares coefficient estimates and the second two columns contain the 2-Stage Least Squares estimates. The numbers in parentheses directly under each coefficient estimate are standard errors.

Table 8. Initial Model Regression Estimates

	OLS Estimates		2SLS Estimates	
VARIABLE	WRAP	HS	WRAP	HS
INTERCEPT	805 (264)	- 2031 (9558)	807 (286)	
i	-1093 (256)	-10770 (3280)	-1096 (266)	
SDUR	- -	28835 (4789)	- -	
BANDS	-	4922 (4141)	-	same as
WGT	.137 (.053)	2.1 (.87)	.137 (.053)	OLS
NODES	-	474 (528)	-	
HS	.185 (.009)	- -	.184 (.011)	
adj R²	.96	.65	.95	.65

The model has many appealing characteristics, especially its linear simplicity. (Although the "true" model may have some non-linear form, the linear model is easy to estimate and interpret and is probably a good approximation to the "true" model.) Also, because we have used a pooled database of two WBS elements, R&D and Production/System Activation, we have constructed a model that forecasts costs for two phases of the life-cycle.

On the other hand, the sample size is small, even with pooling, resulting in a limited number of variables in the regression and causing a moderate amount of collinearity between independent variables. However, this collinearity does not cause coefficient sign changes or sign vacillation.

A few standard errors in the HS equation were high partly because of the small sample size and, possibly, the effects of collinearity. We examined the data for numerical instability in the normal equation solution. Although the coefficients were a little unsteady, as measured by the behavior of sequential parameter estimates, and the ratios of singular values indicated slight ill-conditioning of the X prime X matrix in the HS equation, (with ill-conditioning less apparent in the WRAP equation), there were no sign reversals during sequential entry of variables. Our analysis based on singular values indicates that collinearity is not seriously affecting the stability of the coefficient estimates in the HS equation.

To get deeper insight into the numerical stability of the cross products matrix inverse, we present its singular values in Table 9.

Table 9. Singular Value Decomposition

WRAPAROUND COST EQUATION				VARIANCE PROPORTIONS		
NUMBER	EIGENVALUE	CONDITION NUMBER	INTERCEPT	i	HS	WGT
1	2.762298	1.000000	0.0236	0.0316	0.0333	0.0424
2	0.726683	1.949678	0.0002	0.2885	0.3176	0.0008
3	0.376642	2.708138	0.0106	0.1468	0.2003	0.8002
4	0.134377	4.533913	0.9655	0.5332	0.4488	0.1566
			1.0000	1.0000	1.0000	1.0000

HARDWARE/SOFTWARE COST EQUATION					VARIANCE PROPORTIONS			
NUMBER	EIGEN-VALUE	CONDI-TION NUMBER	INTER-CEPT	i	NODE-INST	WGT	SDUR	BANDS
1	4.071688	1.000000	0.0017	0.0187	0.0086	0.0114	0.0101	0.0033
2	0.892184	2.136289	0.0001	0.0046	0.0000	0.0549	0.7098	0.0000
3	0.474884	2.928151	0.0002	0.6833	0.0171	0.1745	0.0347	0.0015
4	0.344031	3.440239	0.0066	0.2726	0.0056	0.2801	0.1632	0.0650
5	0.197277	4.543060	0.0017	0.0063	0.5944	0.1499	0.0771	0.0598
6	0.019936	14.291317	0.9896	0.0145	0.3744	0.3293	0.0051	0.8704
			1.0000	1.0000	1.0000	1.0000	1.0000	1.0000

The lack of two or more high (> .9) variance proportions in Table 9 indicates that collinearity is not seriously affecting the numerical stability of the least squares solution in either equation, with the possible exception of BANDS and INTERCEPT in the HS equation. Our conclusion is that collinearity is not a serious problem in either equation.

We visually inspected the residuals plots for systematic variation and found that the residuals displayed patterns suggesting the existence of ignored explanatory variables, and/or, the presence of heteroskedasticity. Because heteroskedasticity causes only loss of efficiency and not bias, we have not transformed the model to cancel it. The pattern of non-constant variance is somewhat typical of cross-sectional data and we are not surprised that it is there. Visual inspection of residuals is somewhat subjective and more work, such as regressions if the absolute values of the residuals upon explanatory variables, needs to be done.

In analyzing the residuals of the CER, we note the existence of some outliers. In the wraparound cost equation, systems five and six have residuals that are more than two standard deviations from zero. These residuals are associated with forecasts of R&D costs. Production/System Activation forecasts for systems five and six are accurately forecasted. In the hardware/software equation, system five has outlier residuals for both R&D and Production/System Activation costs.

The hat matrix diagonal was analyzed to determine the influence of particular observations upon the model prediction and on the values of the estimated coefficients. Using the 2p/n cutoff criterion for evaluating the relative size of values along the hat matrix diagonal (where p is the number of independent variables including the intercept an n is the sample size), we see that three systems, five, eleven, and twelve, have high leverage. As was observed in the studentized residuals, we observe that system five has a significant effect upon the regression. The dfbetas for weight (WGT) in the Wraparound equation [1] for R&D costs show that if the R&D observation for system five were to be dropped from the regression data, the WGT coefficient would fall in value and the coefficient on HS (hardware/software costs) would rise. The effect of system five's elimination on the WGT coefficient is the reverse in the HS equation [2]. While system twelve acts like system five in regard to the coefficient on WGT for R&D costs, system eleven acts on WGT in the opposite direction. Even so, the dfbetas of system eleven and twelve are small in comparison to that of system five, making it important to discover (at a later time) why system five has such a significant influence.

The figures below depict the fit of the model. The 45 degree line on each graph is the locus on which the points would fall if the model R-squares were 100%.

Figure 1
R&D Wraparound Costs Per Year
Predicted versus Actual

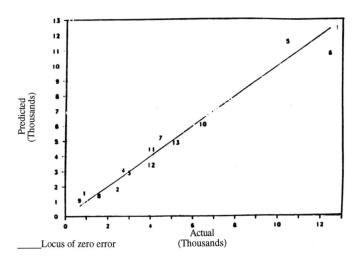

_____ Locus of zero error

Figure 2
R&D Hard/Soft Costs Per Year
Predicted versus Actual

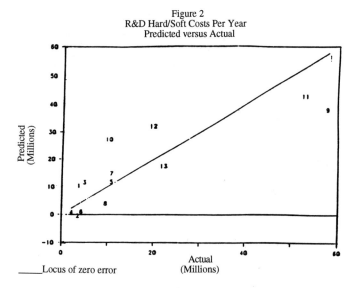

_____ Locus of zero error

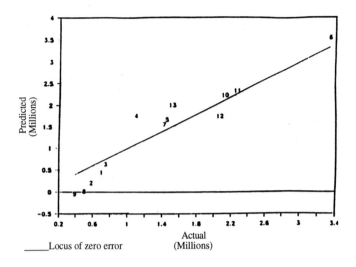

Figure 3
ProdAct Wraparound Costs Per Year
Predicted versus Actual

_____Locus of zero error

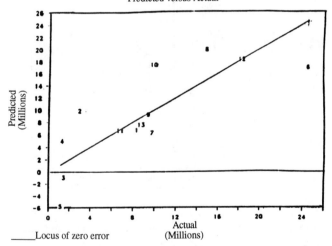

Figure 4
ProdAct Hard/Soft Costs Per Year
Predicted versus Actual

_____Locus of zero error

Additional systems, namely, TACINTEL, MLS, SDI, and MK-15 were added to the analysis as they became available, increasing the size grew from twenty-six observations to thirty-four. Some observations were eventually dropped due to missing data problems.

Ordinarily, increases in the sample size in a regression experiment will improve the accuracy of the coefficient estimates. This assumes, of course, that the new observations are obtained from the same population. The accurate model describing the space-based systems is, however, weakened by the addition of the new systems. The standard errors are much higher (see Table 10), the sign on WGT is reversed in the equation for wraparound (WRAP), and the R^2 is much lower in both equations. Table 10 displays and compares the coefficient estimate before and after the inclusion of the new systems.

Table 10. Re-estimated Coefficients

VARIABLE	OLS Estimates		2SLS Estimates	
	WRAP	HS	WRAP	HS
INTERCEPT	46998 (35458)	-792288 (298241)	27754 (47248)	
i	-48810 (43088)	202486 (142209)	-87636 (58983)	
SDUR	- -	-61815 (220337)	- -	SAME AS
BANDS	- -	262197 (104721)	- -	OLS
WGT	-2.270 (9.752)	17.75 35.80	3.114 (.114)	
NODES	- -	46700 (17024)	- -	
HS	.2353 (.048)	- -	?	
adj R^2	.41	.22	.21	.22

NOTE: Standard errors are in parentheses.

The increase in the standard errors of the extended model shown in Table 10 may be due to the scale of the costs of the added systems and the possibility that the marginal effects of the explanatory variables behave differently in the added systems. Therefore, an intercept dummy (named "SG1", for the 13 original systems in "Systems Group 1"), was added to account for the shift in the intercept term and hopefully re-orient the model to reproduce the same signs and standard errors as in the initial model. Slope effects dummies were not added because the sample size is too small, even with pooling.

The effect of SG1 can been seen in Table 11. The sign on WGT in the WRAP equation reverted to positive and the OLS R-square rises somewhat due to the addition of the SG1 dummy. The equation remains, however, less pleasing than the result obtained with the 13 initial systems.

Table 11. The Coefficients with the SG1 Dummy Added

VARIABLE	OLS Estimates		2SLS Estimates	
	WRAP	HS	WRAP	HS
INTERCEPT	128393 (52305)	−368331 (472443)	−12161 (209519)	
SG1	−124540 (61229)	−308395 (267489)	56551 (287126)	
i	−34798 (41506)	210757 (141531)	−99964 (91214)	
SDUR	− −	20848 (230443)	− −	SAME AS
BANDS	− −	211288 (113167)	− −	OLS
WGT	5.65 (10.04)	31.2 (37.4)	.345 (20.287)	
NODES	− −	28957 (22872)	− −	
HS	.1834 (.0522)	− −	.5115 (.3175)	
adj R^2	.46	.22	.21	.22
NOTE: Standard errors are in parentheses.				

To assess the effect of adding the five non-SG1 systems more closely, the following tactic was used. Instead of adding variables in a stepwise fashion to a regression model drawing upon all available data points, the SG1 model variables were held in the equations, and the additional data points were added in a stepwise manner.

The seven regressions calculated by adding each of the seven data points did not cause a breakdown in the model. The coefficients of both the wraparound equation and the hardware/software equation were insignificantly different to the models that used SG1-only data in their samples. This is probably due to the fact that the 26 SG1 data points outweigh the one non-SG1 datapoint in each regression.

However, when the six regressions consisting of two new data points added to the regression were examined, it was noted that most models suffered from higher standard errors and sign reversals. This may indicate that the addition of the new observations has affected the regression with collinearity; or, more likely, that the new data comes from a population different from that of the SG1 data. For instance, it may not be a coincidence that the coefficient for system weight (WGT) suffered the most when the new data was added. In the SG1 population, weight must be a prime factor because of the fact that heavier systems cost more to launch into space. Non-SG1 systems, on the other hand, may not be as affected by weight and, therefore, the use of weight in a regression model incorporating the non-SG1 systems introduces specification error. In other words, the non-SG1 systems are probably affected by a different combination of factors than those governing the SG1 systems, and it is likely that more and different technical data needs to be collected on non-SG1 cost factors. This is not an indication that extending the SG1-specific model to include other systems is not feasible. It means that we must alter the functional form of the equations to accommodate the new systems by studying the functional differences between them.

SUMMARY AND CONCLUSIONS

We collected and normalized data for thirteen space and communication systems and by pooling R&D and Production/System Activation cost, were able to overcome the small sample size.

We hypothesized that the effects of system complexity transmit their effect down through the life-cycle. We, therefore, developed a two equation system of wraparound and hardware/software costs in which the dependent variable in the second was an explanatory variable in the first. The minimum R-square improvement technique was used to search many alternative models of varying size, and a promising model was identified, with the following qualities:

- The costs addressed are hardware/software cost as a function of system complexity, and integration cost as a function of hardware/software cost and system complexity.

- Each equation predicts integration and hardware/software costs for two components of the work breakdown structure, namely, R&D costs, and Production/System Activation Costs.

- The model is not seriously affected by collinearity, as measured by an examination of the eigensystem of the cross-products matrix and by an examination of singular value decomposition of the variances

- The 2SLS estimate differs insignificantly from the OLS estimate.

- The addition of the five new systems tends to subvert the result. The reasons for this are unknown, and more work needs to be done to discover a successful formulation that includes the data for all systems.

REFERENCES

1.   "Air Force Systems Command (AFSC) Cost Estimating Handbook" (1987), (2 Vols), The Analytic Sciences Corporation, Reading MA.

2.   Belsley, D.A., Kuh, E., Welsch, R. E. (1980), *Regression Diagnostics, Identifying Influential Data and Sources of Collinearity*, Wiley Series in Probability and Mathematical Statistics, Wiley.

3.   Draper, N.R., Smith, H. (1966), *Applied Regression Analysis*, Wiley Series in Probability and Mathematical Statistics, Wiley.

4.   Fischer, Gene H. (1970), *Cost Considerations in Systems Analysis*, RAND Corporation Report, R-590-ASD, American Elsevier, NY.

5.   Johnston, J. (1972), *Econometric Methods*, 2nd edition, McGraw-Hill.

6.   "Military Standard Work Breakdown Structures for Defense Material Items", MIL-STD-881A (1975).

7.   Snedcor, G.W., Cochran, W.G. (1967), *Statistical Methods*, the Iowa State University Press, Ames, Iowa.

8.   Thiel, Henri (1971), *Principles of Econometrics*, A Wiley/Hamilton Publication, Wiley.

9.   Yelle, L. (1979), "The Learning Curve: Historical Review and Comprehensive Survey," *Decision Sciences*, Vol. 10, No. 2, April, pp. 302-328

A Review of Estimate at Completion Research

David S. Christensen, Richard C. Antolini,
and John W. McKinney
Air Force Institute of Technology

Abstract

The cancellation of the Navy's A-12 program has increased interest in forecasting the completed cost of a defense contract, termed "Estimate at Completion" (EAC). In addition, popular software packages and electronic spreadsheets allow users to quickly compute a range of EACs. Analysts and managers are left with the task of deciding which EAC or range of EACs is most accurate. Although there have been many studies that either compare existing EAC formulas and models, or propose new ones, few have been published in journals or magazines, and there is little guidance regarding which formula or model is most accurate. This paper reviews 25 studies which either propose or compare EAC formulas and models. Each study is briefly described. Tables which summarize research results are provided. Results show that no one formula or model is always best. Additional research with regression-based models is needed.

Introduction

On 7 January, 1991, Defense Secretary Cheney announced that the Navy's A-12 program was canceled[1]. Although there were many reasons for the A-12 cancellation (Beach, 1990), certainly the problem of estimating its completed cost was an important contributing factor. Regarding this estimate, Secretary Cheney complained that "no one can tell me exactly how much it will cost to keep [it] going" (Morrison, 1991:30).

In fact, there were many estimates of its cost. Beach reported that the Navy's program manager chose to rely on a lower estimate, despite several higher ones presented by his own analyst. Beach also suggested that "abiding cultural problems" effectively suppressed the more pessimistic estimates. Navy Secretary Garret voiced a similar conclusion. In testimony before the House Armed Services Committee, Secretary Garret dismissed serious errors in judgment by senior career people involved with the A-12 by saying that they were "can-do" people who would not admit failure lightly (Ireland, 1991:27).

Of course, such cultural problems are not unique to the Navy.

---

[1] Technically, the A-12 full-scale development contract was "teminated for default."

Using the same data, Department of Defense, Service, and contractor analysts often disagree about estimated completion costs. Although some of the disagreement may be attributed to cultural bias, the problem of accurately estimating the completed cost of a defense contract remains.

In the last sixteen years, there have been a large number of studies which have explored the problem of estimating the completed cost of defense contracts. Only a few of these "Estimate at Completion" studies have been published in journals or magazines generally available to interested readers. Most are theses, cost research reports, or special studies, and remain "buried" in cost and technical libraries. This paper reviews 25 of these studies, collectively termed "Estimate At Completion Research." Its purpose is to inform the reader of the results of this research, generate insight into the appropriate use of Estimate at Completion (EAC) formulas, and identify areas for additional research.

The paper is divided into three parts. In the first part, EAC formulas are briefly described and categorized. In the second part, noncomparative studies, which advocate or introduce new EAC methodologies, are briefly reviewed and summarized in a table. In the last part, comparative studies, which compare the actual cost of completed contracts against various EAC formulas, are reviewed and summarized in a table. Generalizations based on this review conclude the paper.

EAC Formulas

The EAC can be computed by formula using cost management data provided by the contractor to the Government in the *Cost Performance Report* or the *Cost/Schedule Status Report*. The studies reviewed in this paper assume that data presented in these reports are reliable. The reliability of the data depends upon the degree to which the contractor adheres to a strong system of internal controls involving the scheduling, budgeting, costing, and analysis of contractual effort. See Department of Defense Instruction 5000.2, *Defense Acquisition Management Policies and Procedures*, for a description of these controls.

All EAC formulas are based on the combination of several data elements presented on the cost management report: Budgeted Cost of Work Scheduled (BCWS); Budgeted Cost of Work Performed (BCWP); and Actual Cost of Work Performed (ACWP). These data elements are usually reported monthly. Cumulative and averaged data can then be computed through the period of the contract's life.

For this paper, EAC formulas are classified into three categories: index, regression, and other. The generic index-based formula is shown in Equation 1:

$$EAC = ACWPc + (BAC - BCWPc)/Index \qquad [1]$$

The subscript "c" indicates cumulative data. Budget at Completion (BAC) is the total budget for the identified work. Detailed descriptions of these and other related terminology are presented elsewhere (e.g., Air Force Systems Command Pamphlet 173-4, *Guide to Analysis of Contractor Cost Data*).

The index, normally some combination of ACWP, BCWP, and BCWS, is used to adjust the budgeted cost of the remaining work on the contract (BAC - BCWPc). The assumption implicit in this adjustment is that the contract's past cost and schedule performance is recurrent and reflective of future performance. For this paper, these "performance indices" are classified into four groups:

$$Cost\ Performance\ Index\ (CPI) = BCWP/ACWP \qquad [2]$$

$$Schedule\ Performance\ Index\ (SPI) = BCWP/BCWS \qquad [3]$$

$$Schedule\ Cost\ Index\ (SCI) = SPI\ X\ CPI \qquad [4]$$

$$Composite\ Index = W1\ X\ SPI + W2\ X\ CPI \qquad [5]$$

The weights shown in Equation 5 (W1 and W2) can take on any value from 0 to 1, and normally add to unity.

These indices can be based on monthly, cumulative, or averaged data. For this paper the following labeling conventions are adopted: "CPIm" represents a CPI based on the most recent month; "CPIc" represents a cumulative CPI; "CPIx" represents a CPI averaged over x number of months, beginning with the most recent month and going backwards. For example, CPI3 represents a 3 month average CPI, with the current and the last two previous months included. SPI and SCI use the same conventions. For example, "SCIc" is a cumulative SCI, and "SPI6" is a six month average SPI, with the current and the last five months included.

The indices can be averaged in two ways. Usually, the averaged index is defined as a ratio of sums through x months:

$$CPIx = \Sigma\ BCWPx\ /\ \Sigma\ ACWPx \qquad [6]$$

$$SPIx = \Sigma\ BCWPx\ /\ \Sigma\ BCWSx \qquad [7]$$

An alternative definition is to divide sum of the monthly indices by the appropriate number of months:

$$CPIx = (\Sigma \ CPIm) \ / \ x \qquad \qquad [8]$$

$$SPIx = (\Sigma \ SPIm) \ / \ x \qquad \qquad [9]$$

Unless specified otherwise, this paper defines an averaged index according to Equations 6 and 7.

The second and third categories of EAC formulas are termed "regression" and "other." The regression-based formulas are derived using linear or nonlinear regression analysis. For this paper, nonlinear regression analysis is defined as the analysis of a nonlinear relationship, regardless of whether it can be transformed into a linear relationship.[2] In any case, the dependent variable is usually ACWP, and the independent variable(s) is usually BCWP, a performance index, or time. The "other" category is for any formula that is not in the first two categories, such as formulas based on heuristics.

It is apparent that there are an infinite number of possible EAC formulas. The analyst is left with the interesting task of deciding which formula or group of formulas to use. *Performance Analyzer* (Scifers, 1991), a popular analysis software package, allows the user to chose from a variety of formulas. However, no guidance is provided regarding which formula or group of formulas is most accurate. The remaining parts of this paper will address this issue by reviewing EAC research conducted over the past sixteen years.

Noncomparative Studies

Noncomparative studies do not compare EAC formulas and models. Instead, they describe a "new" formula or forecasting methodology. Generally, each of these studies involves a complicated heuristic or statistical technique that does not lend itself well to comparative analysis. Table I summarizes 13 noncomparative studies by author, year, Service (or sponsoring organization), and forecasting methodology. ("DLA" is Defense Logistics Agency. "DSMC" is Defense Systems Management College.) Several of the studies listed have more than one author. To save space in the table, only the name of the first author is listed. See "References" for a complete listing of authors. A brief description of each study follows:

---

[2] The general linear regression model can be applied to inherently linear models by a suitable transformation of the variables. For example, nonlinear cumulative cost growth patterns, sometimes closely approximated by logistics curves, may be transformed into a linear form before estimating by ordinary least squares.

TABLE I

SUMMARY OF NONCOMPARATIVE EAC RESEARCH

Author (year)	Organization	Forecasting method
El-Sabban (1973)	Air Force	Bayes' theorem
Sincavage (1974)	Army	Time series analysis
Holeman (1975)	DSMC	Performance factor (subjective)
Olsen (1976)	Air Force	Regression/time series analysis
Busse (1977)	Air Force	Nonlinear regression analysis
Weida (1977)	Air Force	Nonlinear regression analysis
Jakowski (c1977)	Navy	Composite index
Parker (1980)	DLA	Composite index (subjective)
Lollar (1980)	Air Force	Composite index
Chacko (1981)	DSMC	Time series analysis
Haydon (1982)	Navy	EAC range analysis
Watkins (1982)	Navy	Time series analysis
Totaro (1987)	DLA	Composite index (subjective)

Index-based methods. Four of the noncomparative studies proposed ways
to develop weights for the composite index. Jakowski (c1977) and
Lollar (1980) suggested formulas for deriving the weights. Parker
(1980) and Totaro (1987) suggested that the weights be subjectively
assigned. Because the SPI is driven to unity at contract completion by
definition, these studies generally suggest that the SPI eventually
looses its information content. Accordingly, the weight assigned to
the SPI should decrease to zero as the contract progresses to
completion. In a fifth study, Haydon (1982) derived a point estimate
from a range of EACs computed by several index-based formulas.

   Jakowski (Navy Aviation Systems Command, c1977) proposed a rather
complicated heuristic for determining the weights of the composite
index. First, CPIc is used until there are significant decreases in
the most recent monthly CPIs. When this happens, an "optimally
weighted" composite index is used. The optimal weighting is defined as
that weight which results in the least historical standard deviation in
the composite index. After the 60% completion point, CPIc is again
used. Original documentation for Jakowski's heuristic could not be
located, but is described by Covach, et al. (1981:24).

   Lollar (Aeronautical Systems Division, 1980) proposed defining the
weights for cumulative SPI and CPI as the relative contribution which
the absolute values of schedule and cost variance percentages make to
their total. Blythe (1982) and Cryer (1986) included Lollar's method
in their comparative studies. It did not do well against the other
formulas.

   Parker's (Defense Logistics Agency, 1980) method consists of
simply computing a range of composite indices, with the weights varying

from 0 to 1 in increments of 0.1. The analyst would then subjectively decide which composite index to be most appropriate given the conditions of the contract.

Totaro (Defense Logistics Agency, 1987) suggested that determining the weights for the composite index be a function of percent complete. Starting weights for the SPI and CPI were subjectively assigned by the analyst after a consideration of program characteristics, such as the manpower loading projected by the contractor.

Haydon and Riether (ManTech Corporation for Navy Weapons Engineering Support Activity (NAVWESA), 1982) proposed a technique to develop a point estimate from a range of EACs computed using various formulas. First, a range of EACs is computed using index-based formulas evaluated by Covach, *et al.* (1981). Second, the range is expanded by 2.5 percent, and the median of this expanded range is taken as the point estimate for the EAC. Based on an analysis of 21 completed or nearly completed contracts (15 development, 6 production) managed by the Navy, if the contractor's EAC was less than this point estimate, the point estimate was the more accurate forecast 79 percent of the time. A sample worksheet for the procedure and a numerical example are provided.

Regression-based methods. Three noncomparative studies proposed using regression analysis to model the curvilinear cumulative cost growth profile typical on defense contracts. As a group, the techniques proposed in these studies are well documented, complicated, and demand considerable knowledge of regression analysis. As such, they would not be easy to implement.

Sincavage (Army Aviation Systems Command, 1974) proposed using time series analysis to forecast the EAC. The computer-based model, "Time Series Analysis for Army Internal Systems Management" (TSARISM), uses moving average, autoregressive, or a combination of the two time series analysis techniques. As such, it is sensitive to the statistical problem of autocorrelation and requires many months of data before it can be developed. Accordingly, the model would only be useful during the later stages of a contract. Based on discussions with the author, the original documentation has been lost.

Olsen, *et al.* (Aeronautical Systems Division, 1976) described a time series forecasting technique used by the B-1 System Program Office. A computer program called "GETSA" developed by General Electric and leased by the B-1 SPO was used to forecast EACs. Other techniques, including regression analysis and exponential smoothing, are also briefly described. A numerical example is provided.

Busse (Air Command and Staff College, 1977) recommended an alternative way to develop coefficients for a nonlinear regression-based model developed by Karsch (1974). Although Busse made no comparisons with the Karsch model, a numerical example based on Karsch data was provided. Comparing the results of Busse with those of Karsch at several contract completion stages indicated that the Karsch model generated more accurate EACs.

Weida (Air Force Academy, 1977) proposed using nonlinear regression analysis to fit development program data to a normalized S-curve. After adjusting the data for inflation and statistical problems (heteroscedasticity and autocorrelation), Weida established that the S-curve was descriptive of cumulative cost growth on each of the 22 development programs which he examined. The normalized S-curve could then be used for both comparative and predictive purposes. A numerical example was provided. Although Weida's technique is complicated, it is compelling and deserves serious attention.

Chacko (Defense Systems Management College, 1981) proposed using a time series forecasting technique termed "adaptive forecasting." According to Chacko, five months of data are necessary before accurate estimates are possible. Essentially, the adaptive forecasting model adapts (changes) as each month's data become available. Accordingly, the model is best suited to short-term forecasting.

Watkins (Navy Postgraduate School, 1982) proposed using linear regression analysis and an adapted form of the Rayleigh-Norden model. According to Watkins, the Rayleigh-Norden model is descriptive of life-cycle patterns of manpower buildup and phaseout on defense contracts. In this study, the model is used in a linear regression analysis of ACWP against time. Quarterly data from three contracts submitting C/SSRs were used in the regression analysis. The data were adjusted for inflation. There was no adjustment for autocorrelation.

Other methods. These noncomparative studies propose forecasting methods which are based on techniques other than regression analysis or performance indices.

El-Sabban (Army Aviation Systems Command, 1973) proposed the use of Bayesian probability theory to calculate an EAC. The method assumes a normal probability distribution, a mean, and a variance for the EAC at the start of the contract. As current data on ACWP become available, the "prior probability distribution" of the EAC is revised using Bayes's formula. Because the model is not dependent on a long history of performance data, it could be especially useful in the early stages of a contract. Overall, the method is clearly presented, although its accuracy was later challenged by Hayes (1977). An example is provided.

Holeman (Defense Systems Management College, 1974) proposed a "performance factor" determined by subjective judgment as a "product improved method" of developing the EAC. Used like a performance index, the performance factor would include a linear combination of variables (contract changes, inflation, schedule variances, overhead fluctuations, technical risk, and cost history). Determining the relative contribution of each is left to the analyst's judgment. Holeman also suggested that a range of EACs should be subjectively determined and evaluated using simulation. A numerical example is provided.

Comparative Studies

Comparative studies compare the predictive accuracy of two or more EAC formulas. The general approach was to collect data on completed or nearly completed contracts, compute EACs using various formulas, and compare each to the reported Cost at Completion (CAC). For studies using a single contract, the comparison was based on deviation from the CAC in dollars; for studies using multiple contracts, the comparison was based on percent deviation from the CAC. Other comparison criteria included the coefficient of correlation (R-squared) and ranking techniques.

Some studies were more thorough than others, and adjusted the data for various problems, such as scope changes, baseline changes, and inflation. In addition, the better studies checked the sensitivity of the results to the stage of completion, the type of weapon system, and the type of contract (production or development).

Twelve comparative studies are summarized in Table II by author, year, Service (Army, Navy, Air Force), contract phase (development, production), and formula/model category (index-based, regression-based). Four subcategories of index-based formulas are presented (CPI, SPI, SCI, Composite). Within each of these, the type of index is listed. The table shows six CPIs (CPIm, CPI3, CPI6, CPI12, CPIc, other), two SPIs (SPIc, other), two SCIs (SCIc, other), and ten composite indices. For the composite indices, the weighting for SPIc is shown to vary from 10 to 90 percent in increments of 10 percent. The "other" category is for any other possibility for a composite index (e.g., a weighting of .75 on a SPI6). Two subcategories of regression models are listed (linear, nonlinear).

The numbers in the columns for development and production contracts indicate the number of contracts of that kind that were used

TABLE II

SUMMARY OF COMPARATIVE EAC RESEARCH

Author (Year)	Contract			CPI						SPI		SCI		Composite										Regre	
	Service	Dev	Prod	CPIm	CPI3	CPI6	CPI12	CPIc	Other	SPIc	Other	SCIc	Other	10	20	30	40	50	60	70	80	90	Other	L	NL
Karsch (1974)	USAF	1						1																	2
Karsch (1974)	USAF		13					1																	2
Heydinger(1977)	USAF	1		1	2			1																1	2
Hayes (1977)	USAF	3	2					1															1		1
Land (1980)	USAF	~10	~10	1	2			1																	2
Covach (1981)	USN	14	3	1	2	2	1	1	3	1		1						1						3	9
Bright (1981)	USA	11			1	1	1	1		1		1	1					1					1	1	1
Blythe (1982)	USAF	7	19					1		1				1	1	1	1	1	1	1	1	1	1		
Price (1985)	USAF	57			1			1															2	2	1
Cryer (1986)	USAF	7	19					1		1				1	1	1	1	1	1	1	1	1	1	1	
Rutledge (1986)	USAF	13	2									1			1										
Riedel (1989)	USAF	16	40	1	1			1				1		1									1		

in the study. The numbers in the formula columns indicate the number
of formulas of that type that were evaluated. For example, Riedel
(1989) evaluated six formulas using data from 16 development and 40
production contracts that were managed by the Air Force. The six
formulas were CPIm, CPI3, CPIc, SCIc, and two composite indices
(.2SPIc+.8CPIc, another weighting).

A brief description of each comparative study follows. The order
is chronological, consistent with Table II.

Karsch (Aeronautical Systems Division, 1974) compared one index-
based formula (CPIc) and two nonlinear models using data from a
development contract managed by the Air Force. In the nonlinear
models, termed "constrained" and "unconstrained," Karsch regressed
ACWPc against BCWPc through 60 months. In the constrained model, one
of the coefficients was held constant; in the unconstrained model, the
coefficient was allowed to vary. The constrained model produced the
most accurate EAC throughout most of the contract's life. Karsch
recommended that production programs be analyzed to establish
generalizability and a range of values for the fixed coefficient in the
constrained model.

Karsch (1976) subsequently evaluated the same formula and models
using 13 production contracts (aircraft and missile) managed by the Air
Force. The constrained model was again the most accurate, for both
aircraft and missile contracts, and for nearly all the life of every
contract examined. Karsch recommended additional research to establish
generalizability. For both studies, sample data were provided.

Heydinger (Space and Missile Systems Organization (SAMSO), 1977)
evaluated seven formulas and models with 42 months of CPR data from one
development contract managed by the Air Force. There were four index-
based formulas (CPIm, CPIc, two versions of CPI3) and three regression-
based models. The two versions of CPI3 were defined as in Equations 6
and 8 of this paper. The regression-based models included the Karsch
constrained model, and two models proposed by SAMSO. Each of the SAMSO
models regressed ACWP and BCWP against time. One assumed linearity;
the other assumed an Erlang equation was descriptive of the
relationship.

Overall, the SAMSO model using the Erlang equation was the most
accurate throughout the contract's life. The Karsch model was more
accurate than the CPI3 equations in the early and late stages of the
contract. Of the index-based formulas, the CPI3 equations were most
accurate. The CPI3 formula that averaged three monthly CPIs (Equation
8) was slightly more accurate than the other CPI3 formula (Equation 6).
Because of the limited sample, the author advised against generalizing
to other contracts and recommended further research.

Hayes (Air Force Institute of Technology, 1977) evaluated one index-based formula (CPIc), a nonlinear regression model (Karsch, 1974), and a modified version of El-Sabban's model (1973) using data from five contracts (three development, two production) managed by the Air Force. Results indicated the Karsch model as most accurate. The modified El-Sabban model was more accurate than the index-based formula (CPIc).

Land and Preston (Air Force Institute of Technology, 1980) evaluated four index-based and two nonlinear regression models using data from 20 aircraft contracts managed by the Air Force. The exact numbers of production and development contracts were not reported. The index-based formulas included CPIm, CPIc, and the two forms of CPI3. The nonlinear regression models evaluated were the "constrained" and "unconstrained" exponential models proposed by Karsch (1980). Overall, the results showed that the index-based formulas were more accurate than the Karsch models, with CPIc the most accurate of the index-based formulas. CPI3, computed as in Equation 6, was slightly more accurate than CPI3, computed as in Equation 8.

Covach, et al., (ManTech Corporation for Navy Weapons Engineering Support Activity, 1981) evaluated 24 formulas and models using data from 17 contracts (14 development, 3 production) managed by the Navy. The formulas included 12 index-based formulas and 12 regression-based models. The CPI-based formulas were CPIm, two CPI3s, two CPI6s, CPI12, CPIc, and three other kinds. Average CPIs were as defined in Equations 6 and 8. The other CPIs involved dividing an averaged CPI into BAC.[3] The two other index-based formulas were SPIc and an unusual use of the SPI, where SPIc is divided into BAC. The 12 regression-based models used ACWPc, BCWPc, or CPIc as the dependent variable, and BCWPc or Time (months) as the independent variable. For each regression, four curvilinear relationships were tested. The SAMSO nonlinear model (Heydinger, 1977) was also considered for evaluation, but rejected because it was too unstable. Unfortunately, the index-based formulas were not compared to the regression-based models.

A summary of the results from comparing index-based formulas is provided in Table III. Average CPIs defined by Equation 6 were generally more accurate than those defined by Equation 8. The equations which involved dividing an averaged index into BAC were

---

[3] Dividing anything other than CPIc into the BAC is an incorrect algebraic simplification of the basic EAC formula presented as Equation 1 in this paper.

completely discredited. Results of comparing the regression-based
models were less clear. No one model always performed well. Once a
model began to perform well, it usually continued to be the best
regression-based model. Finally, for all of the formulas and models
evaluated, EACs computed from level one data in the work breakdown
structure were as accurate as EACs computed at lower levels and summed
to level one.

TABLE III

RESULTS OF EAC COMPARISONS (Covach, et al., 1981)
14 Development and 13 Production Contracts (Navy)

Completion Stage	Best Performing Formulas
Early (0-40%)	CPI3, CPIc, SCIc
Middle (20-80%)	CPI3, CPI6, CPIc, SCIc
Late (60-100%)	CPI3, CPI6, CPI12

Bright and Howard (Army Missile Command, 1981) evaluated 11
formulas and models using data from 11 development contracts managed by
the Army. Nine index-based formulas (CPI3, CPI6, CPI12, CPIc, SPIc,
SCIc, SPIcxCPI6, .5CPIc+.5SPIc, .75CPIc+.25SPIc) and two regression-
based models (one linear, one nonlinear), with ACWP regressed against
CPI, were evaluated at various contract stages.

Summarized results are shown in Table IV. In the early stage,
Bright concluded that the two regression-based models performed better
than the formulas; of the formulas, the composite indices were the most
accurate. The information content of the SPI was shown to decrease, as
composite formulas giving larger weights to SPI were more accurate in
the early stages of the contracts examined. In the middle stages, the
averaged CPIs were most accurate. Bright suggests that when contracts
have significant cost variance growth in the middle stages, an index
averaged over a shorter period is more accurate than one averaged over
a longer period. In the later stages, CPIc and SCI were more accurate.
The SCI was also found to be a reasonably accurate index in the early
stages of the contracts examined. Of various combinations of SCIs
examined, SPIcxCPI6 was the most accurate.

Blythe (Aeronautical Systems Division, 1982) evaluated 12
composite indices using data from 26 (7 development, 19 production)
contracts managed by the Air Force. Weights for the composite indices
varied from 0 to 1, in .1 increments. Blythe's study differed from the
others in that it derived a regression-based model for each index-based
formula. The model was then used to adjust the EAC, usually upward.

Based on this innovative approach, Blythe found that adjusting the contractor's reported EAC was more accurate than any index-based EAC. Of the index-based EACs, weighting the SPIc at .2 was the most accurate at any stage of completion. Blythe made no distinctions between development and production contracts. **Cryer and Balthazor** (1986) subsequently replicated Blythe's study, using the same data and methodology. The results were insensitive to whether the contracts were development or production.

TABLE IV

RESULTS OF EAC COMPARISONS (Bright and Howard, 1981)
11 Development Contracts (Army)

Completion Stage	Best Performing Formula/Model
Early (0-30%)	Regression, Composite, SPIc, SCI
Middle (31-80%)	CPI3, CPI6, CPI12
Late (81-100%)	CPIc, SCI

Price (Air Force Institute of Technology, 1985) evaluated five index-based formulas and one linear regression model using data from 57 development contracts managed by the Air Force. The index-based formulas were CPIm, CPIc, CPI3, and two unusual composite indices. In the first composite formula, the schedule variance percentage (SV%) is multiplied by .75 and added to the cost variance percentage (CV%): 1-CV%+.75SV%. The second composite formula was defined as a weighted combination of three CPIs: .12CPIm+.24CPI3+.64CPIc. Rationale for these formulas was not provided. Results showed CPIc and the first composite formula to be the most accurate, followed by CPI3 and the regression-based model.

Rutledge and DiDinato (Armament Division, 1986) evaluated two index-based formulas using data from 15 contracts (13 development, 2 production) managed by the Air Force. The two formulas (SCIc and .2SPIc+.8CPIc) were evaluated at three completion stages (25%, 50%, and 75%). According to Wallender (1986, p.3), results indicated the composite index to be more accurate than the SCIc. (Wallender briefly described this study; the original study could not be located.)

Riedel and Chance (Aeronautical Systems Division, 1989) evaluated six index-based formulas using data from 56 contracts (16 development, 40 production) managed by the Air Force. The six formulas (CPIm, CPI3, CPIc, SCIc, .2SPIc+.8CPIc, and (X)CPIc+(1-X)SPIc, where X = percent complete) were evaluated at four completion stages (25%, 50%, 75%, 100%). The sensitivity of the results to the type of weapon system (8

aircraft, 5 avionics, and 5 engines) was also evaluated. Generally,
EACs for production contracts were more accurate than EACs for
development contracts. More specific results are summarized in Table
V. The term "PC" stands for the formula using percent complete to
adjust the weights in the composite index. The term "20/80" stands for
a 20 percent weight on the SPIc and a 80 percent weight on the CPIc of
the composite index.

TABLE V

RESULTS OF EAC COMPARISONS (Riedel and Chance, 1989)
16 Development and 40 Production Contracts (Air Force)

| Phase | System | ---- Completion Stage ---- | | | | Overall |
		25%	50%	75%	100%	
Development	Aircraft	SCIc	CPI3	CPI3	20/80	SCIc
Production	Aircraft	SCIc	CPI3	SCIc	CPIc	SCIc
Development	Avionics	SCIc	CPI3	SCIc	CPIc	CPI3
Production	Avionics	20/80	SCIc	20/80	SCIc	20/80
Development	Engine	CPIm	SCIc	CPI3	CPI3	CPI3
Production	Engine	PC	CPIc	SCIc	PC	CPIc

Conclusion

Attempting to generalize from such a diverse set of EAC research
is dangerous. However, the larger and more diverse the number of
contracts used in the study, the more compelling the generalization.
Of the 13 comparative studies reviewed, the number of contracts varied
from one (Karsch, 1974) to 56 (Riedel, 1989) or 57 (Price, 1985), with
Riedel's sample much more diverse than Price's sample. With this
caveat in mind, the following generalizations are provided:

1.   **The accuracy of regression-based models over index-based
formulas has not been established.** Most of the early research in
EAC forecasting (e.g., Karsch, Heydinger, Sincavage, Weida)
involved nonlinear regression or time series analysis, showed
promise, but suffered from small sample sizes. Studies using
larger sample sizes (Land, Bright) had mixed results. Bright
showed a regression model to be more accurate than selected index-
based formulas in the early stages, but suggested that using the
model was not popular because management would not support early,
pessimistic forecasts, however accurate! Despite Bright's
comment, with the wide availability and decreased cost of computer
technology and statistical software, additional research exploring
the potential of regression analysis as a forecasting tool is
badly needed. The innovative and well-documented work by Weida

and Blythe is compelling and worthy of serious attention. In
short, we have the tools and should use them.

2.   **The accuracy of index-based formulas depends on the type of
system, and the stage and phase of the contract.** As detailed in
Tables III, IV, and V, the larger studies (Covach, Bright, Riedel)
document that no one formula is always best.

    a.   Assigning a greater weight to the SPI early in the
contract is appropriate. Because the SPI is driven to unity,
it looses its predictive value as the contract progresses.
SCI-based formulas were thus shown to be better predictors in
the early stages by Covach, Bright, and Riedel. In the late
stages, the SCIc and CPIc have nearly the same values, and
were shown to be accurate predictors by Bright and Riedel.

    b.   The long-asserted (Wallender) accuracy of the composite
index with a 20/80 percent weighting on SPI and CPI,
respectively, is not supported by the evidence. The most
recent and comprehensive study (Riedel) documents the
accuracy on this composite index on only a small subset of
the contracts. Accordingly, the arbitrary use of this
weighting should be avoided. There is no substitute for
familiarity with the contract.

    c.   Averaging over shorter periods (e.g., 3 months) is more
accurate than averaging over longer periods (e.g., 6-12
months), especially during the middle stages of a contract
when costs are often accelerating (Bright, Covach, Riedel).
In addition, computing the average as the "ratio of sums"
(Equations 6,7) rather than as the "average of monthly
indices" (Equations 8,9) results in slightly more accurate
forecasts (Land, Covach).

It is hoped that this comprehensive review will be of value to
analysts and managers involved with EAC forecasting. The use of
Performance Analyzer or other analysis software has reduced the
mathematical burden of developing independent EACs, but it is no
substitute for judgment. In addition, until the "abiding cultural
problems" referenced by Beach are resolved, the accuracy of EAC
forecasting is of secondary importance.

References

1.   Beach, Chester Paul Jr. *A-12 Administrative Inquiry.* Report to
the Secretary of the Navy. Department of the Navy, Washington DC,
28 November 1990.

2.   Blythe, Ardven L. *Validation of ASD/ACCM's Cost Performance*

Analysis Algorithm. ASD Reserve Project 82-135-TUL. Aeronautical Systems Division, Wright-Patterson AFB, Ohio, December 1982.

3. Blythe, Ardven L. *A Stochastic Model for Estimating Total Program Cost*. ASD Reserve Report No. 84-135-TUL. Aeronautical Systems Division, Wright-Patterson AFB, Ohio, undated.

4. Bright, Harold R. and Truman W. Howard, III. *Weapon System Cost Control: Forecasting Contract Completion Costs*, TR-FC-81-1. Comptroller/Cost Analysis Division, US Army Missile Command, Redstone Arsenal, Alabama, September 1981.

5. Busse, Daniel E. *A Cost Performance Forecasting Model*. MS thesis. Air University, Maxwell AFB, Alabama, April 1977 (AD-B019568).

6. Chacko, George K. "Improving Cost and Schedule Controls Through Adaptive Forecasting." *Concepts: The Journal of Defense Systems Acquisition Management* 4:73-96, Winter 1981.

7. Covach, John, Joseph J. Haydon, and Richard O. Reither. *A Study to Determine Indicators and Methods to Compute Estimate at Completion (EAC)*. Virginia: ManTech International Corporation, 30 June 1981.

8. Cryer, James M. and Leo R. Balthazor. *Evaluation of Weighted Indices on Algorithms Utilized for Calculating Independent Estimates at Completion*. ASD/HR MR Project No. 86-225, Wright-Patterson AFB, Ohio, 1986.

9. Department of the Air Force. *Cost/Schedule Control Systems Criteria Joint Implementation Guide*, AFSCP 173-5. Washington DC: HQ AFSC, 1 October 1987.

10. Department of the Air Force. *Guide to Analysis of Contractor Cost Data*, AFSCP 173-4. Washington DC: HQ AFSC, 1 September 1989.

11. Department of Defense. *Defense Acquisition Management Policies and Procedures*, DODI 5000.2. Washington DC, 23 February 1991.

12. El-Sabban, Zaki M. *Forecast of Cost/Schedule Status Utilizing Cost Performance Reports of the Cost/Schedule Control Systems Criteria: A Bayesian Approach*. U.S. Army Aviation Systems Command, St. Louis Missouri, January 1973 (AD-754576).

13. Haydon, Joseph J. and Richard O. Reither. *Methods of Estimating Contract Cost at Completion*. ManTech International Corporation: Virginia, 31 January 1982.

14. Hayes, Richard A. *An Evaluation of a Bayesian Approach to Compute Estimates At Completion for Weapon System Programs*. MS thesis, AFIT/GSM/77Q-21. School of Systems and Logistics, Air Force Institute of Technology (AU), Wright-Patterson AFB, Ohio, September 1977 (AD-A056502).

15. Heydinger, Gerard N. *Space and Missile Systems Organization Cost Performance Forecasting Study*. Cost Analysis Division, Los Angeles California, June 1977.

16. Holeman, J.B. *A Product Improved Method for Developing A Program Management Office Estimated Cost at Completion*. Defense Systems Management School, Fort Belvoir, Virginia, January 1975 (AD-A007125).

17. Ireland, Andy. "The A-12 Development Contract: A Blueprint for Disaster." Remarks to the Institute of Cost Analysis Washington Area Chapter, 12 December 1990. *Newsletter of the Society of Cost Estimating Analysis*, October 1991, pp. 26-27.

18. Karsch, Arthur O. *A Cost Performance Forecasting Concept and Model*. Cost Research Report No. 117. Aeronautical System Division, Wright-Patterson AFB Ohio, November 1974.

19. Karsch, Arthur O. *A Production Study Sequel to the Cost Performance Forecasting Concept and Model*. Cost Research Report 132. Aeronautical Systems Division, Wright-Patterson AFB Ohio, August 1976.

20. Land, Thomas J. and Edward L. Preston. *A Comparative Analysis of Two Cost Performance Forecasting Models*. MS thesis, LSSR 23-80. School of Systems and Logistics, Air Force Institute of Technology, Wright-Patterson AFB, Ohio, June 1980 (AD-A087500).

21. Lollar, James L. *Cost Performance Analysis Program for Use on Hand-Held Programmable Calculators*. Cost Research Report 141. Aeronautical Systems Division, Wright-Patterson AFB, Ohio, April 1980.

22. McKinney, John W. *Estimate At Completion Research - A Review and Evaluation*. MS thesis, GCA/LSY/91S-6. School of Systems and Logistics, Air Force Institute of Technology, Wright-Patterson AFB, Ohio, September 1991.

23. Morrison, David C. "Deep-Sixing the A-12." *Government Executive*, March 1991, pp. 30-35.

24. Olsen, David and Roger W. Ellsworth. *Forecasting Techniques Employed in a Line Organization*. Cost Research Report 127. Aeronautical Systems Division, Wright-Patterson AFB Ohio, February 1976.

25. Parker, Charles W. *C/SCSC and C/SSR Cost Performance Analysis Programs*. DCAS Residency Harris, DCASMS, Orlando, Florida, 20 October 1980.

26. Price, James B. *An Evaluation of CPRA Estimate at Completion Techniques Based Upon AFWAL Cost/Schedule Control Systems Criteria Data*. MS thesis, AFIT/LSY/GSM/855-28. School of Systems and Logistics, Air Force Institute of Technology (AU), Wright-Patterson AFB Ohio, September 1985 (AD-A162282).

27. Riedel, Mark A. and Jamie L. Chance. *Estimates at Completion (EAC): A Guide to Their Calculation and Application for Aircraft, Avionics, and Engine Programs*. Aeronautical Systems Division, Wright-Patterson AFB Ohio, August 1989.

28. Rutledge, Brian and Phil DiDinato. "Estimate at Completion." Letter (signed by Wayne L. Foster), Headquarters Air Force Systems Command, Directorate of Cost Analysis, Comptroller, Headquarters Armament Division, Elgin AFB, Florida, 23 January 1986.

29. Scifers, Thomas. *Software User's Manual for Performance Analyzer Version 3.0*. Camp Springs, Maryland: TRW, Inc., 15 January 1991.

30. Sincavage, John T. *Time Series Analysis for the Army Internal Systems Management (TSARISM).* Cost Analysis Division. Army Aviation Systems Command, St. Louis, Missouri, 3 October 1974.

31. Totaro, Jeffrey A. "A Logical Approach to Estimate at Completion Formulas." *Program Manager* 16:29-33, November-December 1987.

32. Wallender, Timothy J. *HQ Air Force Systems Command Estimate at Completion Formula Justification.* DCS/Comptroller, HQ Air Force Systems Command, Andrews AFB, Maryland, February 1986.

33. Watkins, Harry. *An Application of Rayleigh Curve Theory To Contract Cost Estimation and Control.* MS thesis. Naval Postgraduate School, Monterey, California, March 1982.

34. Weida, William J. *A General Technique for R&D Cost Forecasting,* USAFA-TR-77-12. US Air Force Academy, Colorado, September 1977 (ADA046105).

A CER Approach to Estimating Aircraft Integration Costs

William Richardson
The Analytic Sciences Corporation, 2555 University Boulevard,
Fairborn, OH 45324

ABSTRACT

In the wake of the reduced threat in Europe, President Bush has promised significant reductions in the size of our armed forces (and DoD budgets) and to continue the development of high technology avionics subsystems. As they have in the past, future budget constraints will inevitably mean a further decrease in the number of new aircraft acquisition programs. Future challenges for the cost estimating community will shift from estimating the cost of developing and producing new aircraft to integrating new technology into existing aircraft. This paper presents the results of four CER studies developed for estimating the cost of integrating new avionics subsytems into existing A–10 aircraft.

The first study developed CERs for the following three cost elements: (1) integration engineering; (2) Group A Kit recurring production; and (3) Group A Kit nonrecurring production. Each of these CERs is, in reality, a summation of eight different weight driven CERs. The study is documented in Section A.

Section B describes how installation costs, the subject of the second study, were estimated as a function of modification complexity (as defined by the ELSIE (ELectronic Subsystem Integration Estimator) Model. The CER was the result of regression analysis on previous attack and fighter aircraft case histories.

Kitproof and Trial Installation labor (the third study) were estimated as a function of Installation labor costs. This third CER study is presented in Section C.

The fourth and final study is discussed in Section D. It expressed all other integration cost elements as a percentage factor of the Group A and B kit costs. The factors were based on 10 previous A–10 modification case histories.

SECTION A — GROUP A MODIFICATION COST ESTIMATING RELATIONSHIPS

1. INTRODUCTION

Contents — This section details the cost estimating methodology for the Group A kit cost estimating relationships (CERs). The CER's are in the form of cost to weight relationships for the applicable A–10 airframe structure groups (fuselage structure, wing structure, empennage structure) and aircraft subsystems (electrical, environmental, hydraulics, controls and displays, avionics). Each group and subsystem is in–turn detailed at the functional cost element level, i.e., non–recurring design engineering, non–recurring tooling, and recurring manufacturing costs.

Basis — The foundation for this methodology is a research project performed by the RAND Corporation undertaken in cooperation with the directorate of Cost Analysis, Aeronautical Systems Division, Air Force Systems Command, and the major aircraft contractors in the United States (Ref. 1). The RAND research report provides an estimating procedure, a cost database, and a set of estimating equations which are appropriate to support estimating the modification costs of A-10 aircraft or any other aircraft in the DoD inventory. Included in the report's database are the major functional element costs and man-hours for aircraft structural groups and subsystems for most DoD aircraft designed, manufactured, and modified in the period of 1960 through 1981 including the A-6, A-7, A-10, F-4, F-14, F-15, and F-16.

Summary of RAND Findings — Using the detailed cost data collected from industry RAND was able to develop a set of parametric equations (i.e., CERs), wherein weight is the explanatory, or *independent*, variable and cost is the *dependent* variable. The equations developed by RAND for a given subsystem take the following form: functional element hours equals the subsystem weight (raised to a fractional power) times a constant, whereby the answer represents the cumulative average hours over the first 100 units including the related development hours.

In their search for explanatory variables RAND found, from a statistically significant standpoint, only weight and speed were dependable predictors of cost. Further, speed was significant in only one instance: design engineering for fuselage structures.

For some subsystems/functional cost elements statistically defensible equations based on regression analysis could not be found. In those instances (e.g., design engineering and production labor for the environmental subsystem) the cost data was plotted and estimator judgment was used to determine the best fit curve through the data.

From the historical cost data RAND used regression analysis to develop the cost-quantity improvement curve slope associated with each equation. The improvement curve slope is important because with it the first unit cost can be generated. In the case of the design engineering and tooling functions the first unit cost encompasses the non-recurring design and development costs. For the manufacturing related functions (i.e., production labor and production material) the first unit cost is the starting point from which the manufacturing costs of any and all manufactured quantities can be projected.

2. CER DEVELOPMENT PROCEDURES

The following six tables and related discussion present the development of the Group A kit CERs for each of the subsystems and structure groups likely to be impacted while modifying an A-10 aircraft for enhanced capabilities.

Integration Engineering — Table A-1 captures the derivation of the non-recurring Integration Engineering function CERs for each of eight aircraft subsystems or structure groups. The derivation starts with the RAND equation for the cumulative average engineering labor hours at unit 100 shown in Table A-1, Col. 2).

The following procedures were used:

- Step 1 — Determine the cumulative average hours at unit 100 (see Table A–1, Col. 5) by substituting the appropriate independent variable values (i.e., weight and, in one case, speed) into the RAND equations shown in the second column of Table A–1. The weights (see Table A–1, Col. 4) and speed values were obtained from A–10 aircraft specifications.

- Step 2 — Calculate the first unit (T_1) hours (see Table A–1, Col. 6) by substituting the answer from step 1 (i.e., Y) into the standard improvement curve equation:

$$Y = (T_1) * (100^b) \quad \text{(see Table A–1, Col. 3 for b values)}$$

and solve for T_1. The results are shown in Col. 6 of Table A–1.

- Step 3 — Calculate the T_1 hours per pound by dividing the T_1 hours (i.e., the result of step 2) by the weight (Col. 4 of Table A–1). The T_1 hours per pound (see Col. 7 of Table A–1) represents the engineering hours required to design one pound of new weight associated with a given subsystem or structure group.

- Step 4 — The first unit dollars per pound (T_1 $/lb) by multiplying the T_1 hours per pound (calculated in step 3) by $70.30, the average aircraft industry labor wrap rate according to the latest ASD/ACC study on wrap rates (Ref. 2). Thus the far right column of the table displays the design engineering dollars per pound for each appropriate subsystem or structure group. The Integration Engineering CERs were applied to the *total* weight added plus *one-half* of the weight removed.

Table A–1   Derivation of Nonrecurring Integration Engineering CERs (FY90$)

(1)	(2)	(3)	(4)	(5)	(6)	(7)	(8)
GROUP/SUBSYSTEM	RAND EQUATION T100, CUM AVG HRS	RAND b VALUE	WEIGHT (POUNDS)	T100C.A. (HOURS)	T1 (HOURS)	T1 (HRS/LB)	T1 ($/LB)
Structure, Fuselage	$0.0005S^{1.29}W^{.95}$	-.88	3242	2373	136,545	42.1	2960
Structure, Wing	$84.89W^{.5}$	-.85	4646	5786	289,999	62.4	4387
Structure, Empennage	$4.57W^{.83}$	-.85	958	1363	68,305	71.3	5012
Electrical	$21.4W^{.71}$	-.79	700 *	2241	85,199	121.7	8556
Environmental	Graph Plot Only					180.7 +	12,703
Hydraulics	Not Available					180.7 +	12,703
Controls/Displays	Not Available					126 +	8,858
Avionics	Graph Plot Only	-.86				131 ** ***	9,209

Notes:

W	=	weight in pounds
S	=	speed of 389 knots
*	=	used rough average of fighter/attack aircraft
**	=	used average of Electrical and Avionics T_1 values
***	=	used midpoint of RAND plot data for fighter/attack aircraft
+	=	used A-6 Program History for environmental subsystem as a substitute
T_1 $/lb	=	T_1 hrs/lb x $70.30/hour (FY90 engineering labor wrap rate) (Ref. 2)

Data Peculiarities — For two subsystems, environmental and avionics, RAND was unable to develop statistically defensible equations. As an alternative, the A-6 hours per pound historical value for the environmental subsystem was used. For the avionics subsystem the mid-point of the fighter/attack aircraft database, as compiled by RAND, was used.

The RAND data for the hydraulics subsystem was combined with the flight controls subsystem and as a result the design engineering costs are inappropriately high. If a modification requires rerouting of the hydraulics lines or components, a more cost realistic value must be used. Please note that for the functional elements of tooling, production labor and material, the RAND equations for the hydraulics subsystem are valid because these manufacturing activities relate almost exclusively to the manufacture and assembly of hydraulic lines and components and are extraneous of the design and integration of the flight control laws into the aircraft. This latter point cannot be made for the design engineering function. The environmental subsystem value of $12,703 per pound was selected as the closest substitute considering the design engineering effort involved.

Finally, the RAND data does not separately capture the costs for cockpit controls and displays. Since these items and their design are avionics and electrical related, the simple average of the electrical subsystem, 121.7 hours per pound, and the avionics subsystem, 131 hours per pound was used to arrive at a CER of 126 hours per pound and $8858 per pound for the controls and displays.

Non–recurring Tooling — Table A–2 captures the derivation of the non–recurring tooling function CERs. The derivation process for the tooling CER mirrors that of the Integration Engineering CER, with one exception. Upon derivation of the T_1 hours per pound value, an adjustment was made in recognition that tooling for minor retrofit and modification efforts does not require the relatively expensive major assembly and final assembly tooling which can represent anywhere from 50% to 95% of tooling costs for some subsystems and structure groups. Considering this, we reduced the RAND tooling cost CERs by 90% to arrive at non–recurring tooling hours per pound values for modification efforts. The hours per pound values were in turn multiplied by the industry average tooling wrap rate to arrive at the tooling dollars per pound values for each subsystem and structure group. These appear in the right hand column of Table A–2.

Table A–2   Derivation of Nonrecurring Tooling CERs (FY90$)

GROUP/SUBSYSTEM	RAND EQUATION T100, CUM AVG HRS	RAND b VALUE	WEIGHT (POUNDS)	T100 (HOURS)	T1 (HOURS)	ADJUSTED FOR MODIFICATION		
						T1 (HRS/LB)	T1 (HRS/LB)	T1 ($/LB)
Structure, Fuselage	193.68W [69]	-.82	3242	22,828	996,471	307.4	30.7	2,284
Structure, Wing	48.62W [70]	-.83	4646	17,938	819,918	176.5	17.7	1,317
Structure, Empennage	28.53W [73]	-.85	958	4,283	214,634	224.0	22.4	1,667
Electrical	Graph Plot Only	-.81	700 *			55.0 **	16.5	1,228
Environmental	16.77W [64]	-.78	117	353	12,817	109.5	32.9	2,448
Hydraulics	6.32W [75]	-.76	500 *	668	22,128	44.3	13.3	990
Controls/Displays	Not Available					45 **	13.5	1,004
Avionics	Graph Plot Only	-.76				35 ***	10.5	781

Notes:
W   =   weight in pounds
*   =   used rough average of fighter/attack aircraft
**   =   used average of Electrical and Avionics T_1 values
***   =   used midpoint of RAND plot data for fighter/attack aircraft
T_1 $/lb   =   adjusted T_1 hrs/lb x $74.40/hour (FY90 tooling wrap rate) (Ref. 2)

Recurring Production — Tables A–3 and A–4 capture the derivation of the recurring production labor and production material CERs, respectively. The derivation process in each instance mirrors that of the Integration Engineering CER discussed earlier. The slopes used in the derivation of the T_1 values were 90% for labor and 95% for material which is consistent with modification history detailed in the ELSIE (ELectronic Subsystem Integration Estimator) Model (Ref. 3). The production material values were derived initially in FY77 dollars, as was the RAND database, and inflated to FY90 dollars through the application of a 2.237 multiplication factor. The factor was compiled from the January 1990 OSD inflation rates and

Table A-3   Derivation of Production Labor CERs
(FY90$)

GROUP/SUBSYSTEM	RAND EQUATION T100, CUM AVG HRS	RAND b VALUE	WEIGHT (POUNDS)	T100 (HOURS)	T1 (HOURS)	T1 (HRS/LB)	T1 ($/LB)
Structure, Fuselage	$32.9W^{.83}$	-.152	3242	26,988	46,409	14.3	897
Structure, Wing	$32.1W^{.77}$	-.152	4646	21,387	36,777	7.9	496
Structure, Empennage	$30.0W^{.77}$	-.152	958	5,926	10,190	10.6	665
Electrical	$26.9W^{.68}$	-.152	700 *	2,314	3,979	5.7	356
Environmental	Graph Plot Only	-.152	117			10.3 **	646
Hydraulics	$17.4W^{.81}$	-.152	500 *	2,671	4,593	9.2	577
Controls/Displays	Not Available	-.152				7.8 ***	490
Avionics	Graph Plot Only	-.152				3.2 **	201

Notes:

W	=	weight in pounds
*	=	used rough average of fighter/attack aircraft
**	=	used average of Electrical and Avionics T_1 values
***	=	used midpoint of RAND plot data for fighter/attack aircraft
T_1 $/lb	=	T_1 hrs/lb x $62.76/hour (FY90 production labor wrap rate) (Ref. A-2)

Table A-4   Derivation of Production Material CERs
(FY90$)

GROUP/SUBSYSTEM	RAND EQUATION T100, CUM AVG HRS	RAND b VALUE	WEIGHT (POUNDS)	FY77$ T100 ($)	FY77$ T1 ($)	FY77$ T1 ($/LB)	FY90$ T1 ($/LB)
Structure, Fuselage	$120.5W^{.89}$	-.074	3242	160,550	209,617	65	145
Structure, Wing	$27.7W^{1.03}$	-.074	4646	165,750	216,407	47	104
Structure, Empennage	$517.2W^{.7}$	-.074	958	63,185	82,496	86	193
Electrical	$858W^{.66}$	-.074	700 *	64,752	84,542	121	270
Environmental	Graph Plot Only	-.074	117	5,850	7,638	65	146 **
Hydraulics	$97.9W^{.94}$	-.074	500 *	31,714	41,406	83	185
Controls/Displays	Not Available	-.074					209 ***
Avionics	Graph Plot Only	-.074				32	73 **

Notes:

W	=	weight in pounds
*	=	used rough average of fighter/attack aircraft
**	=	used average of Electrical and Avionics T_1 values
***	=	used midpoint of RAND plot data for fighter/attack aircraft

Inflation factor to go from FY77$ to FY90$ is 2.237 per DoD Inflation Guidance

guidance for Appropriation 3010, Aircraft Procurement. The results derived in Tables A-3 and A-4 are not used separately but form the basis for the derivation of the total manufacturing costs discussed in the following section.

Manufacturing CERs — Table A-5 captures the derivation of the total manufacturing CERs which are essentially the sum of four components: production labor, production material, kitting cost, and quality assurance labor. The production labor and production material components come directly from Tables A-3 and A-4. The kitting costs, in the form of T_1 values, were estimated as a factor of production labor and production material T_1 dollars per pound. The kitting factor, 31%, is the same as that used to price kits for the F-15 aircraft program and it is documented in a recent Air Force/Contractor forward pricing rate agreement. The fourth component, quality assurance labor, also in the form of T_1 values, was estimated as a

Table A–5   Derivation of Recurring Manufacturing T_1 Costs
(FY90$)

(1)	(2)	(3)	(4)	(5)	(6)	(7)
GROUP/SUBSYSTEM	PRODUCTION LABOR T_1 HRS/LB.	PRODUCTION LABOR T_1 $/LB.	PRODUCTION MATERIAL T_1 $/LB.	KITTING COST T_1 $/LB.	QUALITY LABOR T_1 $/LB.	MANUFACTURING COSTS SUM, COLs 3,4,5&6 T_1 $/LB.
Structure, Fuselage	14.3	897	145	323	107	1472
Structure, Wing	7.9	496	104	186	59	845
Structure, Empennage	10.6	665	193	266	79	1203
Electrical	5.7	358	270	196	43	867
Environmental	10.3	646	146	246	77	1115
Hydraulics	9.2	577	185	236	69	1067
Controls/Displays	7.8	490	209	217	58	974
Avionics	3.2	201	32	72	24	329

Notes:

Col. 3   Production Labor Cost = Production Labor Hours x $62.76/hour (Ref. A-2)

Col. 5   Kitting Cost = 31% of Production Labor And Production Material

Col. 6   Quality Labor Cost = 12% of Production Labor Hours x $62.35/hour (Ref. A-2)

factor of production labor hours. The quality assurance factor, 12%, is representative of the aircraft industry average. The resultant quality assurance labor hours were multiplied by the industry average quality assurance wrap rate of $62.35 per hour (Ref. 2) to arrive at the quality assurance dollars per pound values.

The summation of the T_1 dollars per pound values from each of the four components yields the total manufacturing CERs for each of the subsystems and structural groups. These CERs, in terms of T_1 dollars per pound, were applied only to new *added* weight. Various quantities of Group A kits are estimated by combining the appropriate T_1 values with the required quantities on a cost–quantity improvement curve. A 90 percent improvement curve slope is recommended since the vast majority (over 85 percent) of the manufacturing costs are labor related and the production labor curve used in the derivation process is, itself, on a 90 percent slope.

Table A–6 is simply a summary matrix of the CERs derived and presented on Tables A–1, A–2, and A–5.

SECTION B – INSTALLATION COSTS

1.   CER TYPE

Installation hours were estimated as a function of modification complexity. The Electronic Subsystem Integration Estimator (ELSIE) Model method of determining modification complexity was used (Ref. 3).

The ELSIE Model uses a numerical rating process to determine the complexity of a Class V modification. A complexity factor is assigned to each of eight individual integration cost driver categories. The complexity factor is an integer value from zero (least complex) through four (most complex). The

Table A–6   Group A Modification CERs Summary Matrix (FY90$ Per Pound)

| GROUP/SUBSYSTEM | NON-RECURRING | | RECURRING |
	(CES X1100) PRODUCTION LABOR	(CES X1420) TOOLING	(CES X1410) MANUFACTURING T₁ $/LB.*
Structure, Fuselage	$2960	$2284	$1472
Structure, Wing	4387	1317	845
Structure, Empennage	5012	1667	1203
Electrical	8556	1228	867
Environmental	12,703	2448	1115
Hydraulics	12,703	990	1067
Controls/Displays	8,858	1004	974
Avionics	9,209	781	329

*Recurring manufacturing costs on 90% improvement curve

modification complexity is the sum of the individual complexity factors. A sample worksheet containing complexity factor definitions is included as Fig. B–1.

2.   BASIS FOR CER FORMULATION

Regression Analysis — The installation CER was formulated based on regression analysis conducted on installation data from 13 fighter and attack aircraft Class V modification case histories contained in the ELSIE Model database. Two sets of regressions were run — one set for each of the two independent variables, modification complexity and weight. Modification weight is defined as the total weight added to the aircraft plus the total weight removed from the aircraft. These independent variables were selected for the following reasons:

- Modification complexity is the primary driver of the estimating algorithms in the ELSIE Model. ELSIE was developed by TASC for the Aeronautical Systems Division specifically for estimating Class V aircraft modification costs.

- Weight is almost universally accepted by both government and private industry as a logical predictor of cost. It is one of the primary cost drivers of most parametric cost estimating models available on the market today (e.g., PRICE H and FASTE). In addition, some airframe manufacturers (e.g., McDonnell–Douglas) are contractually required to price out Engineering Change Proposals using weight driven CERs which are approved and audited by the government.

- In addition to being widely acknowledged as logical predictors of cost, complexity and weight are both readily available in the ELSIE Model database.

Each of the independent variables was regressed against the following dependent variables (NOTE: In regression analysis the *independent* variable is used to predict the value of the *dependent* variable.):

- Dollars per aircraft — the average cost to install the Group A and Group B kits in the aircraft (i.e., total installation dollars divided by total number of aircraft).

- Dollars per pound per aircraft — the installation dollars per aircraft divided by the total weight (i.e., weight added plus weight removed) involved in the modification.

- Hours per aircraft — the average labor hours required to install the Group A and Group B kits in the aircraft.

Figure B-1   Complexity Factor Description Matrix

COMPLEXITY FACTOR	STRUCTURE CHANGES	INTEGRATION COST DRIVERS						
		INTERFACES WITHIN THE SUBSYSTEM	INTERFACES WITH OTHER AIRCRAFT SUBSYSTEMS	AIRCRAFT EXISTING EQUIPMENT IMPACT	SIGNIFICANT PARTS COUNT	WEIGHT CHANGE	COMPONENT SEPARATION	RELOCATION
0 (None)	No change required to either primary or secondary structure. No structural change required to accommodate equipment or racks. Form-Fit-Function aircraft mod.	Mod is extremely simple with no significant interfaces between components. No interfaces.	Mod is extremely simple with no interfaces with other aircraft subsystems. No interfaces.	Mod can be installed without disturbing any existing equipment. Existing racks and power are available.	Mod can be accomplished without any significant parts. Only small cables, clips, etc. required.	No weight removed or installed.	Mod is simple with minimal wiring changes and easily routing. Insignificant component separation. Average separation: < 1 ft.	Mod is accomplished with little or no work due to relocation. Component relocation relative to prime LRU. < 1 ft.
1 (Minor)	Minor changes to secondary structure caused by LRU relocation or new LRU installation. No changes to aircraft primary structure.	Mod is simple containing a minimum number of interfaces between components. All interfaces are simple. > 0 < 3	Mod is simpler; however, it does contain a limited number of interfaces with other aircraft subsystems. > 0 < 3	Minor impact caused by the necessity to remove/replace a few LRUs in order to accomplish mod. Some minor changes required to existing equipment to accommodate mod.	Minimum number of significant parts required. All parts are simple and easily installed. > 0 < 3	Minor weight removed or installed. > 0 < 25 lbs	There are a few wiring changes and for those changed, the routing is easy, component separation is minor. Average separation: < 1 ft 5 ft	Minor rework required. Component relocation relative to prime LRU. > 1 ft < 5 ft
2 (Average)	Minor changes to secondary structure and some minor changes required to primary structure.	Average number of interfaces between components in the mod. > 3 < 7 simple or > 2 < 4 complex	Average number of interfaces between mod and other aircraft subsystems. > 3 < 7 simple or > 2 < 4 complex	Several LRUs must be removed and replaced to accomplish mod. Minor mods to existing LRUs and other existing equipment to accomplish mod.	Average number of significant parts required. Most parts are simple and can be installed without difficulty. > 3 < 5	Average weight removed or installed. > 25 lbs < 50 lbs	Mod has several wiring changes with average difficulty in routing. Component separation is average. Average separation: > 5 ft < 10 ft	Average rework required. Component relocation relative to prime LRU. > 5 ft < 10 ft
3 (Significant)	Significant changes required to both primary and secondary structure.	Mod is complex containing a significant number of subsystem interfaces. > 7 < 10 simple or > 5 < 8 complex	Mod is complex and contains significant interfaces with other aircraft subsystems. Many interfaces are complex. > 7 < 10 simple or > 4 < 8 complex	Many LRUs must be moved to accomplish mod and then replaced. Mods to existing equipment must be accomplished prior to reinstalling.	Many significant parts required, some of which are complex and difficult to install. > 5 < 10	Significant weight removed or installed. > 50 lbs < 150 lbs	Many wiring changes with difficult routing. Significant component separation. Average separation: > 10 ft < 15 ft	Significant rework required. Component relocation relative to prime LRU. > 10 ft < 15 ft
4 (Major)	Many changes required to both primary and secondary structure. Frames and stringers will be penetrated and/or modified. Racks/hard points must be manufactured and installed.	Mod is very complex. The number of interfaces between mods is large and most are complex. > 10	Many complex interfaces between the integrated subsystem and other aircraft subsystems. Almost all interfaces are complex and significant. > 10	Large number of LRUs must be removed to accomplish mod. Removal/reinstallation is time consuming and difficult.	Large number of significant parts required, many of which are complex and difficult to install. > 10	Major weight removed or installed. > 150 lbs	Large number of wiring changes with extremely difficult routing. Major relocation of components. Average separation: > 15 ft	Extensive rework required. Component relocation relative to prime LRU. > 15 ft
COMPLEXITY FACTOR RATINGS								
MODIFICATION COMPLEXITY FACTOR (Sum of Complexity Factor Ratings)								

- Hours per pound per aircraft — the installation hours per aircraft divided by the weight (i.e., weight added plus weight removed) involved in the modification.

Regression Data — The raw data upon which the regression runs were based was extracted from case histories contained in the ELSIE Model database and represents actual experience on previous Class V modifications on attack and fighter aircraft. The data is summarized in Table B-1. The case history (see Table B-2 for the cases used) gives the following information:

- Modification complexity
- Total installation dollars

Table B–1   Installation Data for Regression Analysis
(FY85$)

MODIFICATION AND NO.	MODIFICATION COMPLEXITY	TOTAL WEIGHT* (POUNDS)	NO. OF AIRCRAFT	INSTALLATION			
				$ PER AIRCRAFT	$/LB PER AIRCRAFT	HRS/ AIRCRAFT	HRS/LB/ AIRCRAFT
A–7 AAS–35 (2951)	18	814	378	$13,056	$16.0	355	4.6
A–7 ALE–40 (2981)	15	542	357	19,988	36.9	535	1.0
A–7 AAN–118 TACAN (12009)	7	127	402	**	**	24	0.2
A–10 INS (3048)	20	1,490	325	126,342	84.8	2,417	1.6
A–10 AVTR/CTVS (3202)	15	203	484	7,936	39.1	311	1.5
A–10 ILS (3183)	17	133	155	16,561	124.5	370	2.8
A–10 BETA DOT (10341)	15	102	125	10,240	100.4	275	2.7
A–10 ALE–40 (3008)	21	344	144	32,014	23.1	1,420	4.1
F–4 APQ–99/162 (12504)	18	612	309	15,909	26.0	350	0.6
F–4 ALR–46 (2777)	18	85	333	1,480	17.4	217	2.6
F–4 INS (19501)	28	498	100	86,910	174.5	1,740	3.5
F–4 UHF SI (59037)	4	25	1,654	3,908	156.3	5	0.2
F–4 ALR–69 (2952)	23	160	428	4,888	30.5	1,250	7.8

* Weight added plus weight removed
** Installation cost data excluded from case history

Table B–2   Modification Case Histories

MOD. NO.	AIRCRAFT	TITLE
2951	A–7D	AN/AAS–35 Laser Search/Track System (Pave Penny)
2981	A–7D	AN/ALE–40(V) Countermeasures Dispenser System
12009	A–7D	AN/ARN–118(V) Airborne TACAN
3048	A–10	Inertial Navigation System (INS)
3202	A–10	Airborne Video Tape Recorder (AVTR) and Cockpit Television Sensor (CVTS)
3183	A–10	AN/ARN–108 Instrument Landing System (ILS)
10341	A–10	Improved "BETA DOT SAS"
3008	A–10	AN/ALE–40(V) Countermeasures Dispenser System
12504	RF–4C	AN/APQ–99/162 Radar System
2777	F/RF–4C/D/E	AN/ALR–46 Radar Homing and Warning (RHAW) System
1950	F–4G	AN/ARN–101 Inertial/Navigation System (INS)
59037	F/RF–4C/D/E/G	UHF System Improvements
2952	F–4D	AN/ALR–69 Compass Tie

- Total weight involved in the modification (i.e., weight added plus weight removed)
- Number of aircraft modified
- Installation hours per aircraft.

The following parameters were calculated from the given data:

- Installation dollars per aircraft
- Installation dollars per pound per aircraft
- Installation hours per pound per aircraft.

Regression Results — A Lotus 1–2–3 based curve fitting program called "CURVEFIT" was used to fit six different curves to each independent–dependent variable pair. Thus, a total of 48 regression runs were made (2 independent variables x 4 dependent variables x 6 curves). The results are summarized in Table B–3.

The power function ($Y = AX^B$), with modification complexity as the independent variable and installation hours per aircraft as the dependent variable, was selected as the "best fitting" curve because it had the highest Coefficient of Determination ($R^2 = 0.91$) of all 48 regressions (see Table B–3). The

Table B–3   Regression Analysis Coefficients of Determination (R^2)

	$/AC	$/LB/AC	HRS/AC	HRS/LB/AC
Independent Variable (X) = Modification Complexity				
Linear (Y = A + BX)	0.23	0.01	0.49	0.40
Logarithmic (Y = A + BLnX)	0.14	0.07	0.36	0.32
Hyperbola (Y = A + B/X)	0.08	0.16	0.23	0.22
Exponential (Y = AE^{BX})	0.25	0.01	0.84	0.60
Power (Y = AX^B)	0.19	0.06	(0.91)	0.60
Reciprocal (Y = A/(A + BX))	0.05	0.01	0.52	0.66
Independent Variable (X) = Weight				
Linear (Y = A + BX)	0.61	0.02	0.44	0.05
Logarithmic (Y = A + BLnX)	0.39	0.08	0.38	0.01
Hyperbola (Y = A + B/X)	0.13	0.18	0.20	0.04
Exponential (Y = AE^{BX})	0.53	0.01	0.28	0.01
Power (Y = AX^B)	0.57	0.06	0.55	0.02
Reciprocal (Y = A/(A + BX))	0.22	0.01	0.10	0.03

following CER was used to calculate installation hours (the A and B coefficients were calculated by CURVEFIT):

$$Y = 0.056X^{3.1733}$$

(Eq. B–1)

Where,

Y   = installation hours
X   = modification complexity

Note: Multiple regression analyses using weight and modification complexity as the independent variables were also conducted but the R^2 values showed little improvement over the single variable analysis results. The power function, for example, improved only slightly from 0.91 to 0.93. Therefore, the simpler CER was used.

SECTION C — A–10 INTEGRATION FACTORS

1.   INTRODUCTION

This section briefly details the research and development of the A–10 unique aircraft integration factors which were utilized during the course of completing the A–10 Night Attack Avionics and Mission Enhancement Analysis.

2. DATA COLLECTION

The key to this research effort as in any research effort was the availability of A–10 unique historical electronic subsystem integration data. Due to the project's aggressive schedule and the limited availability of data at WPAFB it was decided to limit the data search to the Electronic Subsystem Integration Estimator (ELSIE) project available at TASC. A survey of all available ELSIE historical integration program case histories produced a list of fifteen candidate A–10 subsystem integration programs to be investigated (see Table C–1). The ELSIE case histories provide program unique cost, schedule, and technical information, where available, from the following documents:

Table C–1   ELSIE A–10 Aircraft Case Histories

	EQUIPMENT NOMENCLATURE/DESCRIPTION	MODIFICATION #
1.	Inertial Navigation system (INS) ASN–141	3048
2.	Airborne Video Tape Recorder (AVTR)	3202
3.	AN/ABN–108 Instrument Landing System (ILS)	3183
4.	BETA DOT Stability Augmented System (SAS)	10341
5.	AN/ALE–40 Countermeasures Dispenser System	3008
6.	Ground Collision Avoidance System (GCAS)	3301
7.	Continuously Computed Impact Point (CCIP) Gunsight	3293
8.	Turbine Engine Monitor System (TEMS)	11308B
9.	Electro-Luminescent Formation Lights	30098B
10.	Self Defense Air Missile AIM–9	3232
11.	AN/ALE–40 Countermeasures Dispenser System Improvement	13614
12.	APN–232 Combined Altitudes Radar Altimeter (CARA)	10611
13.	CPU–142A Central Air Data Computer (CADC)	41652
14.	AAS–35 Pave Penny	13405
15.	Flight Control Clearance Improvement	10342

- Program Management Directive (PMD)
- Modification Proposal and Analysis (MPA)/Fm48
- Cost Performance Report (CPR)
- Contract Funds Status Report (CFSR)
- Contractor Cost Data Report (CCDR)
- Time Compliance Technical Order (TCTO)
- Acquisition Plan (AP)
- Program Management Plan (PMP)
- Technical Data Sheet
- Contract
- Jane's Aircraft and Avionics Books.

In addition to the Program Management Directive (PMD), other key documents utilized in the development of the A–10 integration factors are identified and defined below:

- Modification Proposal Analysis (MPA) — A comprehensive technical study and cost and schedule analysis that considers all aspects of a proposed Class V mod. A Class V mod is one that either provides a new or improved operational capability or enhanced operational safety. The PMD directs either Air Force Systems command (AFSC), AFLC, or a joint team to prepare an MPA.

- Fm 48 — Basically the same (except for format) as an MPA, but used for Class IV mods. Class IV mods correct material deficiencies to ensure safety, correct reliability and maintainability deficiencies, and correct electromagnetic compatibility or communications security deficiencies. The main difference concerning this study was that Group A and Group B kit costs were not broken out within the Fm48, but were on the MPA.

- Time Compliance Technical Order (TCTO) — Contains instructions, list of material, size and weight information, and the labor hours for kit installation on a specific MDS aircraft. Prepared by the system manager (SM) at the ALC.

3.   ANALYSIS OF DATA

The data within each ELSIE A–10 subsystem integration program case history was analyzed for completeness and reconciled against the other source documents, where applicable, to ensure consistency. Again, the three primary documents for each mod program are the PMD, MPA/Fm48 and the TCTO. Cost data normally appears in a PMD only after submission of an MPA/Fm48. Adjustments to these costs by USAF are common. Therefore, the cost data appearing in the latest document, usually the PMD, were used during data analysis. Installation hours appear in both the MPA/Fm48 and the TCTO. Since the TCTO is based on actual trial installation data, installation hours identified in the TCTO were used to recompute installation costs. Note: During the review and analysis of the A–10 case histories it was found that five of the cases (Nos. 11 through 15 in Table C–1) provided incomplete/insufficient cost/technical data for factor development and were omitted.

The detailed cost elements in both the MPA and Fm48 documents were grouped by cost category using the MIL–STD–881A aircraft system Work Breakdown Structure (WBS) tailored for specific use on modification programs. Table C–2 details the fourteen WBS cost categories and allocation by source document utilized in the ELSIE analysis. Table C–3 details the sixteen WBS cost categories utilized for the A–10 analyses. Note: In most cases information was simply transferred from one table to the other or involved the combining/merging of ELSIE database WBS elements into the A–10 WBS format (see Table C–3).

Table C–2   ELSIE Cost Categorization

WBS COST CATEGORY	COST ELEMENT	
	CCBIR AFLC FORM 48	MPA AF FORMS 2612, 2613, 2614
Development	—	Development
Non–Recurring	Engineering Engineering Data Changes Trial Installation Software Engineering Proofing Special Installation Tooling	Engineering (Group A) Prototype Testing Proofing Tools (Production/Installation) Computer Programs
Group A Kit	Kits and Materials	Mod Kit Cost
Group B Kit		Mod Kit cost
Aircraft Installation	Installation Labor	Installation Cost (Contract/Depot)
Training Equipment	Training Equipment	Maintenance/Crew Trainers
Training Equipment Installation	—	Installation Cost (Contract/Depot)
Peculiar Support Equipment (PSE)	Peculiar Support Equipment	Support Equipment (Peculiar)
PSE Installation	—	Installation Cost (Contract/Depot)
Common Support Equipment	—	Support Equipment (Common)
Investment Spares	Initial Spares (Investment)	Initial Spares Cost Investment Spares Cost Investment M&O Parts Cost Expense
Data	TCTOs Technical Data Manual Changes Tapes/Cards	Data Provisioning Data
Other	—	Bench Test Sets Mod of In Stock Spares Mod of Components Equipment Rental Contractor Logistics Support

¹Configuration Control Board Item Record

Table C–3   A–10 Cost Categorization

WBS NUMBER	WBS COST CATEGORY	ELSIE DATABASE COST ELEMENT
11100	Integration Engineering	Engr (Group A & B), engr (Trng equip), engr (supt equip), prototype testing, trial installation, Kit proofing
11120	Engineering Change Orders	None
11300	Data/Manuals	Group A, Group B, Trng equip, supt equip, provisioning, reprocurement
11400 11410 11420	Group A Kits Material (Recurring) Non–recurring	Group A Kit None
11500 11510 11520	Group B Kits Material (Recurring) Non–recurring	Group B, Group B mod kits, None
11600	Peculiar Support Equipment	New euip, mod equip, mod depot supt, ground equip, Engr
11700	Trainers	CPT mod kit, MTS mod kit, TMSA mod kit
11800	Tooling	Tooling (Group A & B)
11900	Ind Validation/Verification	None
12000	Interim Contractor Support	None
12100	Flight Testing	None
12200	Other Labor (Installation)	Aircraft Installation
12300 12310 12320	Initial Spares Investment Expense	Group A, Group B, Trng equip, supt equip, mod of spares Group A, Group B, Trng equip, supt equip, WRSK/BLSS
12400	Software	Engr (software)
12500	RDT&E	None
12600	Common Support Equipment	None

4. COST FACTOR DEVELOPMENT

The A–10 modification case histories are presented and summarized in Table C–4. The A–10 integration factors were then derived from the cost database provided in Table C–4 as a percentage of the total costs for Group A and Group B kits (see Table C–5) and of Integration Engineering (see Table C–6). The factors are presented by program and cost category and include a composite "simple" and "weighted" average cost factor for each cost factor based upon a summary of the ten programs. These factors may be used for estimating A–10 unique subsystem integration costs for those cost categories for which the analysis was done.

Table C–4 A–10 Modification Case Histories ($K)

WBS NUMBER	WBS ELEMENT DESCRIPTION	INS #3048	AVTR #3202	ARN-108 #3183	BETA DOT #10341	ALE-40 #3008	GCAS #3301	CCIP #3293	TEMS #11308B	LIGHTS #30098B	AIM-9 #3232	TOTAL
	Modification Quantity	410	486	157	176	146	655	655	650	665	653	4,653
11100	Integration Engineering	$8,707	$869	$1,526	$806	$2,758	$7,688	$1,197	$1,672	$3	$4,604	$29,830
11200	Engineering Change Orders											
11300	Data/Manuals	5,332	358	137	184	129	2,900	1,118	1,000	3	3,000	$14,161
11400	Group A Kits	50,576	5,894	3,221	5,893	22,502	12,177	655	0	0	5,208	106,126
11500	Group B Kits	156,618	7,010	3,514	9,953	5,259	22,711	3,930	78,600	1,147	19,530	308,272
11600	Peculiar Support Equipment	9,353	88	206	881	850	3,240	430	15,700		3,658	34,316
11700	Trainers	4,054	4	0	98	0	527	839	2,028	0	363	7,913
11800	Tooling	324	0	0	0	0	0	0	0	0	0	324
11900	Ind Validation/Verification											
12000	Interim Contractor Support											
12100	Flight Testing											
12200	Other Labor (Installation)	41,061	3,841	2,567	1,280	4,610	14,760	1,691	6,100	3,060	1,306	78,585
12300	Initial Spares	11,712	0	24	0	785	7,326	1,691	2,998	125	2,927	27,588
12310	Investment	8,581		24		785	5,456	1,446	2,798	125	1,950	
12320	Expense	3,131					1,870	245	200		977	
12400	SSoftware				48		1,172					1,220
12500	RDT&E											0
12600	Common Support Equipment											
	Total Cost	287,737	18,064	11,195	19,143	36,893	72,501	9,860	108,098	4,338	40,506	608,335
	Summary A-10 Modification Data:											
	Average Installation Cost ($K)	$100.15	$7.90	$16.35	$7.27	$31.58	$22.53	$0.00	$9.38	$4.60	$2.00	$16.89
	Total Integration Engineering ($K)	8,707	869	1,526	806	2,758	7,688	1,197	1,672	3	4,604	29,830
	Total Group A & Group B Kits ($K)	207,194	12,904	6,735	15,846	27,761	34,888	4,585	78,600	1,147	24,738	414,398

Table C-5   A-10 Modification Costs as a Factor of Group A & Group B Kits

SYSTEM MODIFICATION	INTEG ENG	DATA	PSE	TRAINING	TOOLING	INSTALL	SPARES	SOFTWARE
INS	4.2%	2.6%	4.5%	2.0%	0.2%	19.8%	5.7%	0.0%
AVTR	6.7%	2.8%	0.7%	0.0%	0.0%	29.8%	0.0%	0.0%
AN/ARN-108	22.7%	2.0%	3.1%	0.0%	0.0%	38.1%	0.4%	0.0%
BETA DOT	5.1%	1.2%	5.6%	0.6%	0.2%	8.1%	0.0%	0.3%
AN/ALE-40	9.9%	0.5%	3.1%	0.0%	0.0%	16.6%	2.8%	0.0%
GCAS	22.0%	8.3%	9.3%	1.5%	0.0%	42.3%	21.0%	3.4%
CCIP	26.1%	24.4%	9.4%	18.3%	0.0%	0.0%	36.9%	0.0%
TEMS	2.1%	1.3%	20.0%	2.6%	0.0%	7.8%	3.8%	0.0%
FORM LIGHTS	0.3%	0.3%	0.0%	0.0%	0.0%	266.8%	10.9%	0.0%
AIM-9	18.6%	12.1%	14.4%	1.5%	0.0%	5.3%	11.8%	0.0%
Simple Average	11.8%	5.5%	7.0%	2.6%	0.0%	43.5%	9.3%	0.4%
Weighted Average	7.2%	3.4%	8.3%	1.9%	0.1%	19.0%	6.7%	0.3%

Table C-6   A-10 Modification Costs as a Factor of Integration Engineering

SYSTEM MODIFICATION	INTEG ENG	DATA	PSE	TRAINING	TOOLING	INSTALL	SPARES	SOFTWARE
INS	N/A	61.2%	107.4%	46.6%	3.7%	471.6%	134.5%	0.0%
AVTR	N/A	41.2%	10.1%	0.5%	0.5%	0.0%	0.0%	0.0%
AN/ARN-108	N/A	9.0%	13.5%	0.0%	0.0%	168.2%	1.6%	0.0%
BETA DOT	N/A	22.8%	109.3%	12.2%	0.0%	158.8%	0.0%	6.0%
AN/ALE-40	N/A	4.7%	30.8%	0.0%	0.0%	167.2%	28.5%	0.0%
GCAS	N/A	37.7%	42.1%	6.9%	0.0%	192.0%	95.3%	15.2%
CCIP	N/A	93.4%	35.9%	70.1%	0.0%	0.0%	141.3%	0.0%
TEMS	N/A	59.8%	939.0%	121.3%	0.0%	364.8%	364.8%	179.3%
FORM LIGHTS	N/A	100.0%	0.0%	0.0%	0.0%	N/A	4166.7%	0.05
AIM-9	N/A	65.2%	77.5%	7.9%	0.0%	28.4%	63.6%	0.0%
Simple Average		49.5%	136.6%	26.5%	0.4%	155.1%	499.6%	20.1%
Weighted Average		47.5%	115.0%	26.5%	1.1%	263.4%	92.5%	4.1%

5.   COST FACTOR APPLICATION

The cost factors developed in the course of this analysis may be used to develop credible cost estimates for aircraft subsystem installations and modifications on the A-10 aircraft. However, care should be exercised in the selection of the appropriate factor to be utilized. Program unique factors may be used directly or adjusted based upon expert opinion should one of the ten programs evaluated work as an analogous system. The average factors may also be used in the event none of the systems evaluated are analogous to the planned subsystem integration/modification. The cost analyst would be well advised to review the case histories thoroughly and fully understand the breadth and depth of each A-10 modification before a final factor is selected. The raw data used to develop the factors are available in the ELSIE final documentation and/or the ELSIE case history files.

SECTION D — KITPROOF AND TRIAL INSTALLATION COSTS

1.   CER TYPE

Kitproof labor hours and Trial Installation labor hours were estimated as a function of aircraft kit installation hours.

2.   BASIS OF CER FORMULATION

The CERs were based on the ratio of kitproof hours or trial installation hours to aircraft kit installation hours. Historical data from the five A–10 Class V modification case histories contained in the Electronic Subsystem Integration Estimator (ELSIE) Model database (Ref. 3) was used to formulate the CERs.

The raw data and the arithmetic computations used to derive the CERs are summarized in Table D–1. (Note: Table D–2 provides more complete information regarding the identity of the case histories used.)

Table D–1   Kitproof and Trial Installation

(1) MOD	(2) INSTALLATION HOURS*	(3) KIT PROOF HOURS*	(4) RATIO (3 P 2)	(5) TRIAL INSTALL HOURS*	(6) RATIO (5 P 2)
INS	2,417	8,580	3.55	12,890	5.33
AVTR	311	352	1.13	528	1.70
ILS	370	392	1.06	588	1.59
BETA DOT SAS	275	560	2.04	840	3.05
AN/ALE–40	1,420	2,800	1.97	10,800	7.61
		Total	9.75	Total	19.28
			P5		P5
		Average	1.95	Average	3.86
		Rounded	2.0	Rounded	4.0

*Average hours per aircraft

Table D–2   Modification Case Histories

MOD. NO.	AIRCRAFT	TITLE
3048	A–10	Inertial Navigation System (INS)
3202	A–10	Airborne Video Tape Recorder (AVTR) and Cockpit Television Sensor (CVTS)
3183	A–10	AN/ARN–108 Instrument Landing System (ILS)
10341	A–10	Improved "BETA DOT SAS"
3008	A–10	AN/ALE–40(V) Countermeasures Dispenser System

3. COST ESTIMATING RELATIONSHIPS

Historically, kitproof hours for A–10 modification programs are approximately twice as great as the average kit installation hours per aircraft and trial installation labor is four times as great (see Table D–1). Therefore, the following CERs were used:

Kitproof hours = 2.0 x installation hours

Trial Installation hours = 4.0 x installation hours

4. CONCLUDING REMARKS

The four CER studies presented in this paper collectively constitute the framework for the development of a model for estimating the aircraft integration costs associated with new avionics systems. Though developed specifically for the A–10 aircraft, the techniques employed are universal and could easily be applied to a different set of data to develop a similar set of CERs for any aircraft. Nonetheless, the reader is cautioned that the CERs were derived from an A–10 database and that application to other aircraft in their current form would be a misuse of their intended purpose.

REFERENCES

1.   Large, J.P., et. al., "A Method for Estimating the Cost of Aircraft Structural Modifications," *R–2565–AF*, RAND Corporation, Santa Monica, California, 1981.

2.   Pitstick, J.L., "Aerospace Industry Wrap Rate Survey," ASD/ACC, Wright–Patterson AFB, Ohio, August 1986.

3.   Goldberg, R.L., Electronic Subsystem Integration Estimator, *TR–5300–1*, TASC, Fairborn, Ohio, June, 1987.

V. Cost and Production Analysis

Integrating Cost Analysis and Production System Analysis Using Rapid Modeling Technology

Gregory Diehl
Network Dynamics, Inc., 128 Wheeler Road, Burlington, MA 01803

Rajan Suri
Department of Industrial Engineering, University of Wisconsin,
1513 University Avenue, Madison, WI 53706

ABSTRACT

The ability to forecast product costs and total costs as a function of changes in production mix, volumes, processes and strategy is a fundamental objective of cost analysis in manufacturing. The need to understand the production implications, which in turn impact the cost implications of changes, follows directly. Using a Rapid Modelling Technology (RMT) approach to estimate the factory changes we can easily generate the underlying production data needed for a complete and systemic cost analysis. The decision paradigm starts by building a baseline factory and cost model. The analysis proceeds by comparing this "complete model" with other potential "complete models". Through the use of a few examples we demonstrate the analysis method and its generality. Examples include i) overtime decisions, ii) make versus buy decisions and iii) implications of component quality.

INTRODUCTION

The basis of manufacturing decision making is "can I increase profits"? This question is then translated into a host of other more detailed questions. In the past the answers to most of these questions have been answered from within accounting systems. We propose to add tools and techniques that provide additional information about changes to a manufacturing system and when combined with other data provide better answers to the questions asked.

Initially we outline the need for cost analysis and some unresolved issues within current decision methods. We then propose the integration of factory modeling tools with financial data and justify the integration through a simple example. The production model is introduced and a few interesting points are discussed. The financial model is then described.

Three examples illustrate the usefulness of the approach. The first example is a management decision about the appropriate level of overtime. There is a trade-off between increased cost of overtime, with decreases in Lead Time and WIP, and less overtime leading to higher WIP and Lead times. The second example involves a factory deciding about the level of fabrication within the plant as opposed to purchasing a fully completed component from a supplier. The impact upon the factory of completely out-sourcing the component is reviewed. The third example is a decision concerning the effects of changing to a supplier with more reliable components, albeit at a higher cost. The question to be answered is "does the decrease in rework and scrap materials justify the increased material costs?"

In all three examples we illustrate the need for a manufacturing model so as to be able to calculate the financial impact of the decisions. The complete manufacturing model also identifies a number of changes which are not obvious and usually go unquantified. A side benefit is the ability of the decision maker to understand the non-cost effects (e.g. Lead Time and slack capacity) of different decisions.

NEED FOR COST ANALYSIS

Many of the basic decisions made by manufacturing companies impact their abilities to do business, their costs and their plant capabilities and capacities. For example, basic questions such as "what, where and how should I produce my goods/services" require extensive information about costs, capacities and the business' capabilities. Other questions such as "should I make or buy certain components", "how do I allocate the resources available" and "what is the most profitable level of production" are also of importance to the plant manager and support staff. The quality of these decisions is an important determinant in the success of the business [Hayes 1980].

ISSUES IN CURRENT METHODS

Current financial accounting systems are able to track the costs of production as they occur and also perform allocations of indirect and overhead costs to individual products. Through the use of computers, software, bookkeepers and accountants, companies are able to track each item of their costs as closely as they wish. These data are necessary for producing financial statements as required by stockholders and various government agencies. However, in their raw form, these data are not the information that managers need in order to make decisions about *changes* in their manufacturing systems. This is self-evident from papers in the academic literature: Kaplan [Kaplan 1988] "One Cost System isn't Enough", Cooper and Kaplan [Cooper 1987] "How Cost Accounting Systematically Distorts Product Costs" etc.

Additional problems can arise when a manufacturing concern uses the available cost accounting procedures. For example, by using different, but widely accepted, cost allocation methods, a company can be led to make different decisions on which products are profitable and thus should be continued, and which are unprofitable and should be discontinued [HBS 1985]. Furthermore, radical changes in a manufacturing system can invalidate a cost model[1]. For example, adding numerically controlled equipment may reduce the need for direct labor, increase other costs and open opportunities for new products. A cost allocation scheme based upon direct labor hours is no longer meaningful in this new factory[2]. In this situation it is difficult for a manufacturer to decide whether to implement a radical change [Burstein 1984] , [Cooper 1988].

PROPOSAL OF INTEGRATION OF RMT AND COST ANALYSIS

We propose the creation of a cost/benefit analysis methodology for manufacturing businesses, as distinct from financial accounting, cost tracking or cost control systems. The opportunity that we see is an easy to use, simple to understand, quick, "what-if" oriented tool which subsumes a cost analysis system and a comprehensive manufacturing model (see figure 1). It will be able to fulfill the need

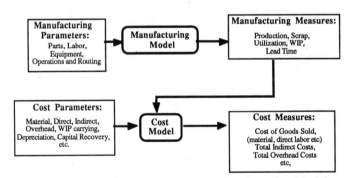

Figure 1: Cost analysis coupled with Manufacturing Model

for a good analysis tool for manufacturing decision support covering such issues as make/buy, quality/cost tradeoffs, cost of lead time reduction, steps needed to minimize

---

1 Even if a new cost model is created for the new manufacturing system it may be difficult to compare this new model to the previous one.

2 The impact of changes of the level of direct labor upon total costs is very different in a highly automated factory from the effects of changes in a factory where the costs are dominated by direct labor expenses.

WIP, impacts of new technology and training as well as other cost related production decisions.

WHY INTEGRATION IS NECESSARY

When one uses only the cost numbers and does not also focus on the underlying manufacturing system as well, the following systemic errors can be introduced: 1) non-linear effects in costs (e.g. fixed costs) 2) capacity constraints (e.g. utilization cannot exceed 100%) 3) inter-product interferences and efficiencies (e.g. shared components) 4) nonlinear effects in manufacturing (e.g. WIP versus production graph, see figure 2). These are discussed below.

Non-Linear Effects in Costs

Non-linear effects in costs occur when increasing production by one unit causes a sizeable jump in the resources required. A simple example is the need for a machine to be able to produce parts. This purchase occurs for the first piece but a second machine does not have to be purchased until some much higher level of production. This "lumpiness" is non-linear and must be included.

Capacity Constraints

In some decision models, capacity constraints are not considered or a long run constraint is used in place of a more realistic and smaller constraint. For example, no machine is really ever 100% busy (or if it is 100% busy, the WIP grows without bound), some time, however small, is lost due to a lack of materials, labor or for maintenance.

Inter-Product Interferences and Efficiencies

Inter-product efficiencies and interferences are ubiquitous. As production is switched from one product to another, the need for a change over time or setup represents a "product interference". The ability of two main products to use a shared component leads to less chance of component shortage, decreased purchasing costs and usually to decreased engineering costs, thus providing an "inter-product efficiency".

Non-Linear Effects in Manufacturing

In a real manufacturing system, very few items are related in a linear fashion. For example, work-in-process (WIP) levels are almost never linear with production. An $x\%$ increase in production leads to greater than $x\%$ increase in WIP (see figure 2).

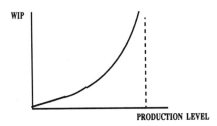

Figure 2. WIP versus Production Level

Even the simple relationship between piece production and equipment utilization may not be linear because as the production level increases a manufacturer could also change the number of lots run and the size of the lots.

A simple example of an Integrated Circuit fabrication line shows why it is especially beneficial to use a manufacturing model when generating a cost model. The labor and capital costs and production levels are fixed during the month we are considering. The only choice to be made is the lot size. Most cost accounting systems will assign costs to parts based upon the amount of time they are using the equipment and labor resources. Some will allocate the cost for idle resources and some will not. However, it is generally accepted that if a part increases the resources it uses then that part should cost more. Also it is generally believed that if a part increases its use of resources then other parts suffer some negative consequence. The numbers below outline a case, quite common in practice, where decreasing the lot size of one part decreases its WIP, and the WIP levels for all other products simultaneously.

Part Name	Production Level	Setup Time	Run Time	BASELINE Lotsize	WIP	NEW SYSTEM Lotsize	WIP
P1	100	2	1	10	5.0	5	4.1
P2	200	2	1	10	10.0	10	9.8

Table 1: Comparison of WIP changes caused by Lotsize changes

In a cost accounting system which has no manufacturing model this change is not included. More significantly, the part with the decreased lot size, will "cost more", even though it creates the benefits for all. This is because more setups are occurring and thus P1 is using more of the available resources. In actuality, since the equipment and labor costs are fixed they should not be considered in the analysis of the best lot sizes. Only the WIP carrying cost should be considered. A model of the manufacturing system is required to determine how the WIP will change.

MANUFACTURING MODELS

The types of models that are able to estimate overall manufacturing system performance are few in number. Basically two technologies are currently used, discrete event Monte-Carlo simulation and queueing network theory. Each method has its advantages and disadvantages, both in theory and in practice [NDI 1987] , [Suri 1991]. In comparison to other possible manufacturing models they are able to estimate or calculate 1) good and scrap production levels, 2) equipment utilization 3) dynamic resource interferences (i.e. Multiple Man Resource issues (MMR)) and 4) WIP and lead times. The technologies are able to incorporate real life complications such as equipment breakdowns, randomness in the delivery of raw materials and variability in actual process times; all of which affect the performance measures just listed [NDI 1991].

We believe that a manufacturing model based upon queueing network theory has distinct advantages over simulation technology [Solberg 1977] , [Suri 1991]. All of the capabilities and features we perceive as important are available within this framework. These include those listed previously (e.g. production levels, resource utilization, MMR, WIP, Lead Time, equipment breakdowns, and variability) as well as the speed of calculation, ability to create models without programming and the sufficient scope of the problems that can be addressed by the technology. For example, we can model the effects of shifting to a Just-In-Time strategy (JIT) by 1) decreasing the production lot size, 2) decreasing the variability in deliveries, 3) decreasing the variability from the standard process time, 4) decreasing machine setup times and 5) decreasing equipment repair times [World Bank 1991]. These capabilities have been demonstrated through a number of commercially available software packages [NDI 1991].

The first step in the process of creating a model is to build or gather the necessary production related information. This includes the number and type of labor and equipment resources available, the expected levels of production and the indented bill of material (IBOM) or product assembly structure. Also required is information about how the part is routed through the production process and the required process and setup times at each step. This represents enough information to build a model of the manufacturing system.

An example of the information needed and results given by a manufacturing model follows in eqs 1-3 below[3]:

Model Input Information:
P_i =   Number of pieces of product i to be made per month

---

[3] This model does not include labor, equipment failures, variability in arrivals and service times. These features can be added but complicate the model without illustrating the points made here.

L_i = Lot size for product i

S_{ijk} = Setup time per lot of product i, manufacturing step j,
at machine k

R_{ijk} = Run time per piece of product i, manufacturing step j,
at machine k

A_k = Available time per month for machine k

Model Output Information:

W_{ijk} = Work in Process for Product i, manufacturing step j,
at machine k

L_{ijk} = Lead Time (time spent) for a lot of Product i, manufacturing
step j, at machine k

U_k = Utilization of machine k

Model Calculations:
  from definitions

$$U_k \quad = \quad \sum \; [\; P_i * R_{ijk} + (P_i / L_i) * S_{ijk}\;]\; / \; A_k \qquad \text{(eq 1)}$$

[Utilization = Total of all work / Available Time]

and from Little's Law [Kleinrock 1975]

$$W_{ijk} \quad = \quad P_i * L_{ijk} \qquad\qquad \text{(eq 2)}$$

*[Work-In-Process = Production Rate * Lead Time]*

and estimating

$$L_{ijk} \quad = \quad [\; L_i * R_{ijk} + S_{ijk}\;]\; + \sum \{\; W_{mjk} * (L_m * R_{mjk} + S_{mjk})\; \}$$

$$\text{for all products m} \qquad \text{(eq 3)}$$

Lead Time = Time to setup and run lot + Queue Time
= Time to setup and run lot + Expected number of jobs ahead
*in queue * (setup and run time for those lots)*

From this simple manufacturing model we can generate the previous example and illustrate useful relationships. We created a graph of WIP versus production rate using this model (see figure 2, previous section). Also we illustrate a useful relationship between lot size and lead time (i.e. why larger lot sizes eventually lead to longer lead times), in figure 3. At lot sizes smaller than optimal, WIP is increasing because the increased number of lots leads to increased setup time leading to higher utilization and thus more queueing. At lot sizes larger than optimal, WIP is increasing because parts spend more time waiting for the rest of pieces in the lot to be completed before moving to the next operation and from Little's Law (eq 2) increased lead time is equivalent to increased WIP. The illustration in table 1 (see previous section) is a case where the lot size is initially too large and is decreased towards the optimal, with a resultant decrease in WIP.

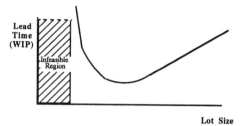

Figure 3: Lead Time (WIP) versus Lot Sizes

PROCEEDING TO COMPLETE THE MODEL

The second step is to identify all non-production related resources. These include, but are not limited to, the engineering staff, the production management, the company management, the plant building, computer and communications equipment, repair staff, facilities and equipment, field service staff and equipment, materials used in production but not shipped, such as lubricants, solvents, etc. One needs to include any item that management may change due to a change in the production of parts.

The final step in completing the model is to provide cost information for all of the items defined so far. The user needs to provide cost information about the material costs, the direct labor costs, equipment depreciation, heat, lights etc, as well as the costs for all of the non-production resources such as plant and company management, engineering staff, etc. The indirect costs are given as raw numbers (e.g. dollars spent in the personnel department) and not as an allocation factor (e.g. overhead dollars per labor hour).

An example of the types of cost information we are seeking is given by adding the cost components to the previously defined manufacturing model:

Let Std_k =   Number of standard hours running each piece of equipment k

L_k =   Number of Labor people running equipment of type k

LD_k =   Direct Labor Cost per Hour for running equipment k for standard time

FL_k =   Fixed Labor Cost per Year per Person (e.g. Health benefits, pension, etc.)

OLC_k =   Labor Cost for running equipment k for one hour of overtime

IL   =   Number of Indirect Labor people

ILC =   Cost per Hour per person for Indirect Labor people each working for the standard time

N_k =   Number of pieces of equipment of type k

FC_k =   Fixed Cost for equipment k running for standard time (Including Electricity, Heat, Plant etc.)

O_k =   Total Number of overtime hours running equipment k

I   =   Interest rate or \$ paid per month for \$1 of WIP

E_i =   Value or Expense of product while in production[4]

WC_i =   WIP carrying cost for product i

Calculating we have

$$WC_i = I * E_i * \sum_{\text{all } j,k} \{ W_{ijk} \}$$

[WCi =   Interest Rate * Value per piece * No. of Pieces in production]

The total cost for the manufacturing system is as follows:

Materials costs $= \sum_{\text{for all } i}$ {Cost of 1 raw material piece of Product i

                 * No. pieces of raw material for i to get 1 good piece

                 * Pi (production level of product i) }

Fixed Equipment Costs = $\sum_{\text{for all } k} \{FC_k * N_k\}$

Direct Labor Costs   = $\sum_{\text{for all } k} \{Std_k * LD_k * L_k\} + \{FL_k * L_k\}$

Indirect Labor Costs   = $\sum \{IL * ILC * Std\}$

Overtime Labor Costs = $\sum_{\text{for all } k} \{ OLC_k * O_k * L_k\,^{5} \}$

WIP costs   = $\sum_{\text{for all } i} \{ WC_i \}$

This cost model provides a portion of the whole model. It contains enough information to create a "Cost of Goods Sold" section within a financial statement along with some pieces of the indirect cost section.

---

[4] This can be a rough estimate or carefully calculated using cost allocation rules. For a more accurate estimate a number of decisions need to be made, which are beyond the scope of this paper. Also, if all products have similar flow times, the inaccuracy in the total WIP carrying expense due to valuation errors is negligible.

[5] We assume that all labor people will work the full amount of overtime.

Note that no cost allocation rules are included. Cooper and Kaplan [Cooper 1991] strongly advocate an approach of not fully allocating all costs to individual parts. We are investigating the extreme case of allocating no costs to individual parts.

EXAMPLES

Using this simple integrated model we now consider three different decisions that frequently occur within manufacturing firms. The first example concerns a choice of the level of overtime to be used so as to gain a decrease in the manufacturing lead time. The trade-off is between more overtime, with lower WIP and lead time and less overtime, with higher WIP and lead time. The second example is a "Make versus Buy" decision where a firm is deciding whether it should buy partially finished components and finish them in-house or whether it should purchase complete components. The problem here is being able to identify the true cost of finishing the component in house. "Should (or which) overhead costs be included?" The third example is a decision about the value of higher quality components. A higher quality component has implications on the level of rework, scrap and resource utilizations. The overall effect of the change needs to be quantified and the change in costs need to be identified.

Overtime Example

In traditional manufacturing firms the usual response to a short-term problem with long lead times is to add short-term capacity (i.e. overtime). We now answer the question "How much overtime will be necessary and how will it cost to decrease the manufacturing lead time by 20%"?

First we create a graph of Lead Time versus Available Time (see figure 4) by running the production model previously described with a number of different levels of overtime (e.g. 0%, 2%, 4%, etc). This graph identifies the tradeoff of overtime and lead time. As overtime is increased, lead time decreases (WIP also decreases proportionally from eq 2). In the example, the manufacturing system is currently running at point (A, B). The 20% decrease in Lead Time (moving from A to A') requires a shift in overtime from B to B'.

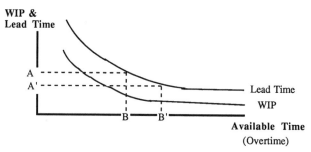

Figure 4: Lead Time and WIP versus Available time

The second step is to create a graph of total costs which also illustrates the tradeoff between WIP costs and overtime costs (see figure 5). An increase in overtime leads to increased overtime cost, but the increase is partially offset by a decrease in the WIP costs.

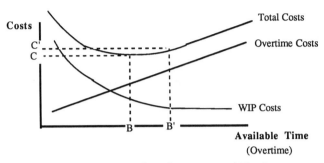

Figure 5: WIP costs, Overtime costs and Total costs

The combination of figure 4 (Lead Time versus Available Time) and figure 5 (Total costs vs Available Time) gives enough information to answer our question. We choose the appropriate lead time numbers on figure 4 (points A and A'), giving the level of available time (i.e. overtime) necessary to achieve them (points B and B'). The total cost curve in figure 5 is then read for the two available time values (i.e. overtime) producing the total cost comparisons (points C and C').

MAKE versus BUY Example

The second example of a MAKE versus BUY decision involves a number of more complex issues. The usual method of doing the cost-benefit analysis is to compare the cost of making the product internally with the cost of the product purchased from an outside vendor. The major complexity is what is the internal cost and whether or not overhead costs should be included within that cost.

Our proposed approach does not directly solve the problem identified within traditional analysis but substitutes a more direct and underlying question. This underlying question is "If the product is not manufactured in-house, what costs (including overhead costs) will disappear and what costs will remain"? Traditional analysis assumes that the costs subsumed within the cost of the part will disappear when the product is purchased from outside the company. This may or may not be true.

To analyze this decision we build two factory models, the first when the product is being built in-house and the second when it is being purchased.

The differences between the two models are given in table 2 below:

CHANGES TO INPUT DATA	MAKE	BUY
Direct Labor (people)	12	8
Direct Labor Overtime (%)	13.3	0
Drill Equipment (no. of pieces)	5	1
Yield at Component Inspection	95%	99%
Next step after failure at TEST	Rework	Scrap
Material Cost for Component	$8	$18

Table 2: Factory Changes between "MAKE"
component and "BUY" component scenarios

The changes to the input data are due to a decrease in the work done in-house leading to a decrease in the direct and indirect labor forces and a decrease in the number of pieces of equipment needed. The number of inspection stations remains the same as the company has decided to inspect the components in the same way it inspected the manufactured pieces before. The cost of the material has increased because of the completed work by the vendor.

The results of the analysis are summarized in the following tables:

CHANGES IN FACTORY PERFORMANCE	MAKE	BUY
Direct Labor Utilization	80.5	79.7
Assemblies in Process	950	899

Table 3: Resulting Factory Differences between MAKE and BUY

CHANGES IN FINANCIAL DATA	MAKE	BUY
Direct Labor Costs (excld. OT)	30720	20480
Direct Labor Overtime Costs	4608	115
Equipment Costs	7877	5452
Total Material Cost	25200	39455
Engineering Costs	2000	2000
Total Factory Costs	70405	67502

Table 4: Cost Differences between a MAKE component and
a BUY (purchased) component

The major differences between the two factories is the decrease in the amount of Drill equipment that is needed, a decrease in direct labor and a decrease in the labor overtime. These lead to lower labor costs at the expense of higher material costs.

There is a decrease in the WIP level because the components are no longer on the floor. There is also a slight increase the the available capacity leading to a shorter lead time for the other parts in the factory. The overall cost of the factory has decreased slightly. This is because the direct and indirect labor costs and overtime costs (the major change) decreased more than the material costs increased.

In a traditional cost analysis an allocation of the the labor and equipment costs to the individual parts is performed. Overtime costs for labor are allocated to all parts. The cost calculated in this way is compared to the cost of the component as purchased. This allocation of cost hides the fact that we can eliminate the use of expensive overtime and substitute the use of standard labor hours instead. See table 5 below for a more explicit comparison of the costs of parts.

Cost of Components $ Per Piece	Current System	Projected Changes	Actual Changes
For Assembly Part			
Direct Labor Costs	12	12	13
Direct Labor Overtime Costs	5	5	0
Equipment Costs	23	23	23
Material Cost (excld. compnt)	3	3	3
Engineering Costs	2	2	2
Total Cost (excld. component)	45	45	41
For Component Part			
Direct Labor Costs	4	1	1
Direct Labor Overtime Costs	1	0	0
Equipment Costs	5	1	1
Material Cost	8	18	18
Engineering Costs	2	2	2
Total Cost of Component	20	22	22
For TOTAL ASSEMBLY Part			
Total Cost	65	67	62

Table 5: Comparisons of Assembly and Component Costing Methods

The first column is the cost of parts for the "MAKE" factory. The second column is the cost of parts in the "BUY" factory as projected by traditional methods. The third column is the actual costs in the "BUY" factory as calculated using traditional methods. The difference in costs is due to the elimination of overtime costs and the resulting changes in the costs of all parts, not just the cost of the component. Because the traditional analysis fails to predict the third column correctly it suggests the factory should continue to make the component rather than buy it.

Component Quality Example

The third example shows how a change in component quality can have wide effects across a manufacturing system. The process of analyzing the dollar benefits of the changes is complex and thus partially explains why many companies do not understand the importance of quality.

A manufacturer of printed circuit boards is offered a higher quality product by one of its suppliers. The claim is that it will lead to lower scrap and less rework of the assembled circuit boards. The current problems appear to be excessive lead times and bottlenecks in some of the insertion operations. The capacity in the test and repair area was a problem in the past but that appears to be solved. Scrap is a nagging irritation but everyone has learned to live with it.

A sample of the data for the factory appears below:

Labor Name	No. in Group
Insertion Operators	4
Test & Repair	3
Assembly Person	1

Table 6: Labor List and Number of people

Equipment Name	No. in Group	Labor Assigned
Dual-In-Line insert	3	Insertion Operators
Single-In-Line insert	3	Insertion Operators
Radial insert	3	Insertion Operators
Axial insert	3	Insertion Operators
AGV cart	1	Assembly Person
Wave Solder	1	Insertion Operators
Test	6	Test & Repair
Rework	2	Test & Repair
Assembly	1	Assembly Person

Table 7:Equipment List, Number of pieces and
Labor Assigned to Run Equipment

Part Names	End Demand	Lotsize
Board ABCD.001	46000	75
Board ABCD.002	18500	75
Board ABCD.003	37300	75
Board ABCD.004	54500	75

Table 8: Part Names, Volume to be shipped and Lotsizes

Operation Name	Equipment Name	Setup Time (Mins per Lot)	Run Time (Mins per Piece)
DIP Insert	Dual-In-Line insert	20	2.8
SIP Insert	Single-In-Line insert	38	2.6
Radial Insert	Radial insert	16	2.0
Axial Insert	Axial insert	40	1.4
Move on Cart	AGV cart	4	0
Solder	Wave Solder	0	2.1
Return on Cart	AGV cart	4	0
Test	Test	10	3.6
Rework	Rework	2	6.7

Table 9: Operations for Board ABCD.001

From Operation	To Operation	% of Flow
Test	STOCK	85.%
Test	Rework	15.%
Rework	Test	50.0%
Rework	SCRAP	50.%

Table 10: Routing Through Test and Repair for Board ABCD.001

Indented Bill of Materials - Board ABCD.001		
Assembly Name: Board ABCD.001		
Component: Chip Set		1 unit per assembly

Table 11: Assembly and Component Units per Assembly

The initial analysis by the cost accountants identified the cost of the scrap components and the cost of the pieces that were scrapped because of the bad components as well as the cost of the rework done to salvage some of the pieces.

In the analysis below we identify the changes that would occur within the facility if the new supplier was chosen. The initial changes in the scrap rate would lead to decreased product flow through the test and rework areas and well as through the insertion areas. The decrease in the upstream flow (i.e. in the insertion areas) would be due to a smaller number of pieces needing to be started into the line. In addition, the lot sizes would be decrease, rather than a decrease in the number of lot starts[6]. This decreased flow would lead to shorter queuing times and coupled with smaller lot sizes would result in shorter overall lead times. In table 12 we summarize the differences.

CHANGES IN FACTORY PERFORMANCE	Original Supplier	Improved Supplier	Percentage (%) change
Average Yield at Test Operation	82	96	17 %
Percent of Boards Scrapped	17.3	3.2	81%
Average Utilization at Insertion	81.8	75.7	7.5%
Average No. Boards In Process	1760	1100	37.5%
Average Lead Time for Boards (days)	2.21	1.58	28.5%
No. Chip Sets Required (in 000s)	183	161	12%

Table 12: Changes in factory performance due to
increased component quality

On the cost side of the analysis we identify the total savings, initially excluding the increased cost of the component (See table 13). This allows us to identify the maximum amount that the increased quality is worth. The only changes are in the decrease of material required because of decreased scrap and the decrease WIP carrying cost because of the decreased lead time and WIP. No labor or equipment would be removed from the factory so those expenses will not change. The business value of shorter lead times to the end customer is important, however it is beyond the scope of this analysis and therefore is not quantified here.

CHANGES IN FACTORY COSTS (in 000s)	Original Supplier	Improved Supplier
Material Costs (without premium for quality)	2240	2040
WIP Carrying Costs	21.3	13.2

Table 13: Changes in the factory cost due to
a change in component quality

The observation that there is little other change in the total cost is an instance of a more general statement *"If managers fail to follow up any reductions in the demands on organizational resources, improvements will create excess capacity, not increased profits"* [Cooper 1991, pg 135]. In order to reap the full benefits of a change in supplier a manager must either (i) identify which expenses will change (e.g. eliminating

---

[6] This leads to a shorter lead time rather than a larger decrease in the utilization. See figure 3 previously.

resources and associated costs) or (ii) specify how those extra resources will lead to increased output and thus higher profits.

The value of the increased quality is realized through cost savings and other factors that are external to this analysis. These other factors include such items as value of shorter delivery times to customers, less field service or product returns and speed to market with new technology. The composite model that we have built here is able to identify the manufacturing changes that occurred. This allows the manager to explore the alternatives of decreasing resources and costs, or of increasing the production level or of providing better service and quality.

CONCLUSIONS

The ability to easily change a manufacturing parameter and then to observe the resulting impact on the costs has a profound effect on a manager's ability to understand the operation of his/her business. A manager changes an item under his/her control and can watch the effects in terms of both the manufacturing system as well as in the costs. The coupling of a manufacturing model with a cost model provides this ability. The user-friendliness of software and the removal of the manual tasks (e.g. data entry and transferal) allows decision makers to focus on the data, the decision and other business considerations and not be focused on the effort necessary to get "an answer".

REFERENCES

[Burstein 1984] M. Burstein & M. Talbi, "Economic Justification for the Introduction of Flexible Manufacturing Technology: Traditional Procedures versus a Dynamics-Based Approach" in Proceedings of the First ORSA/TIMS Special Interest Conference: Flexible Manufacturing Systems, Ann Arbor, MI 1984, pp.100-106.

[Cooper 1987] R. Cooper & R.S. Kaplan "How Cost Accounting Systematically Distorts Product Costs", Chapter 8 in William Bruns, & Kaplan (eds) Accounting and Management: Field Study Perspectives (Boston: Harvard Business School Press), 1987

[Cooper 1988] R. Cooper & R.S. Kaplan "Measure Costs Right: Make the Right Decisions", Harvard Business Review vol 64, No. 5 Sept-Oct 1988. pp 96-103

[Cooper 1991] R. Cooper & R.S. Kaplan, "Profit Priorities from Activity-Based Costing", Harvard Business Review Vol 67 No. 3 May-June 1991, pp 130-135.

[Garlid 1988] S. Garlid, B. Fu, C. Falkner & R. Suri, "Evaluating Quality Strategies for CIM Systems", Printed Circuit Assembly 2, 5 (June 1988), 23-27.

[HBS 1985] R. Cooper "CAMELBACK COMMUNICATIONS, INC.", Harvard Business School case study 9-185-179 (1985)

[Hayes 1980] R.H. Hayes & W.J. Abernathy, "Managing our way to economic decline", Harvard Business Review vol 58, No. 4 July-August 1980. pp 67-77

[Karmarkar 1988] U. Karmarkar & S. Kekre "Manufacturing Configuration, Capacity and Mix decisions considering Operational Costs", Journal of Manufacturing Systems, vol 6, no. 4 page 315-324.

[Kaplan 1988] R.S. Kaplan, "One Cost System Isn't Enough", Harvard Business Review Jan-Feb 1988, pp 61-66

[Kleinrock 1975] L. Kleinrock, *Queueing Systems I*, J. Wiley & sons, New York, N.Y., 1975

[NDI 1987]*MANUPLAN II User Manual*, Network Dynamics Inc., Burlington Mass, 1987.

[NDI 1991] *MPX User Manual*, Network Dynamics Inc., Burlington Mass, 1991.

[Pritsker 1986] A.A.B. Pritsker *Introduction to Simulation and SLAM II*, Halsted Press, New York, NY, 1986.

[Solberg 1977] J.J. Solberg "A Mathematical Model of Computerized Manufacturing Systems", Proc. 4th Interl. Conf. Prod. Resrch. Toyko, Japan 1977.

[Suri 1991] R.Suri & S. deTreville "Full Speed Ahead" OR/MS Today May 1991.

[World Bank 1991: 1,2,3] A. Moody, R. Suri, J.L. Sanders, "The Global Challenge of Modern Manufacturing Technology and Practices: (1) Footware Industry (2) Bicycle Industry (3) Printer Circuit Board Industry", World Bank Technical Reports, World Bank, Washington D.C. 1991.

[Zimmerman 1987] J. Zimmerman "Accounting Incentives and the Lot Sizing decision: a field study", Univ of Rochester, working paper CMOM 87-05

Policy Formation for National Industries: The Use of Non-Linear Programming Allocation of Joint Costs

Dennis E. Kroll and Kristjan B. Gardarsson
Department of Industrial Engineering, Bradley University, Peoria, IL 61625

INTRODUCTION

This paper is written with the Icelandic fishing industry in mind, namely the freezing plants. The subject at first is how to allocate joint-cost to finished products. A joint-cost is a cost which incurred up to the split-off-point (S.O.P.), which is the point in the manufacturing process beyond which the individual products are clearly identifiable. The motivation for allocating joint-cost to the individual products in this business like others is the demands of financial and tax reporting to trace all product related costs to finished goods. It is also necessary because the prices are both derived and/or determined from the cost of the goods.

Common methods for allocating joint cost are: proportional to some physical measure of the output (weight), or either the market value of the products immediately after the product is clearly identifiable, or the net realizable value of the finished products. One reason or argument for focusing on allocation of joint-cost within the Icelandic fishing industry is the "value added tax" (VAT), a new reality in Iceland. This kind of tax makes it very important to know and understand where in the manufacturing process values have been added to the products, and how the cost caused by this value adding should be allocated to the finished products in a fair manner.

However, the introduction of a new tax vehicle suggests further discussion of the problem of planning a national industries, especially in a mixed economy. In many smaller and/ or emerging countries, government aid in guiding new industries remains a necessary condition. Government policies affect all industries in all countries. Yet these policies are very political things. A method of at least starting the process of forming policy from a mathematically sound basis is desirable. When considered from this point of view, the use of the proposed technique is of even greater value to industries as well as companies.

The intention here is to demonstrate a technique which makes it possible to obtain simultaneously the optimal price-output decision and a joint-cost allocation that is consistent with this decision. A simple example will be considered, where the focus is set on how to allocate the common raw-material and overhead costs to different types of product outputs in a freezing plant.

First the process of packing fish will briefly be described. The non-linear programming model will then be developed and solved for this typical situation. Finally the interpretation of the solution will be discussed in light of the goals of this paper.

THE PROCESS

The processing line begins with the fish stored in chill rooms on pallets. The fish are then transported to the processing line on a fork lift as they are required by production. The pallets are placed on a lifting table to keep the boxes at the work bench level once an operator has taken off the topmost box. Conveyed on roller conveyors,the boxes filled with fish are taken to the box tipper. The contents of the boxes are emptied into a water filled storage tank and a vertical conveyor takes the fish to the holding tray of the heading machine. The empty box is knocked from the tipper by the box following behind, is piled on a pallet, and finally taken to the box washing machine.

The operator at the heading machine takes the fish from the holding tray and inserts it into the heading machine. The headless fish drops into the holding tray of the filleting machine. Here the operator lifts the fish and puts it into this machine where it is filleted and skinned. The next process step is the manual trimming and cutting of the skinned fillets. Here some bones are removed from the fillets and also parts from fins, bones, black membrane and skin which is leftover by the passage through the filleting process are cut off. Also parts which are bloodshot are cut off. The trimming process can be finished by dividing the fillets to go into various types of packages, i.e. "5 lbs" or a "Block". After the trimming operation, the fillets are divided into boneless fillets, waste and white cut off. The waste is transported to the waste container, while the fillets and the white cut off are weighed.

The fillets undergo a quality inspection; the single portions are packed manually into packages which are specially made for the respective products. The products are placed in freezing frames and put into the horizontal plate-freezer. After freezing, the blocks or packs are pushed out of the freezing frame and packed in a film or in cartons as required. For the production of single fillets the bottoms of the storage trays can be removed and the fillets taken to the gyrofreeze spiral belt freezer on a conveyor belt directly beneath the trays. The frozen fillets are wrapped in a film, individually labelled, gathered into collective packs, and placed on pallets which go into the cold store [4,5].

THE ALLOCATION MODEL

Any product produced by a manufacturing process which produces multiple products that adds significantly to the total market value is called a joint-product. A joint-cost is the cost which is incurred up to the split-off-point, which is the point in the manufacturing process where the individual products become clearly identifiable. The joint-cost can be raw materials, direct labor cost and/or some overhead costs such as indirect materials, indirect labor and other indirect factory costs. The split-off-point within this kind of manufacturing process would be at the trimming process.

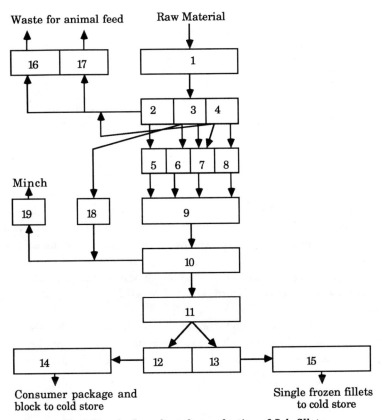

Fig.1 Simple flow-chart for production of fish fillets.

1) Washing-machine	10) Trimming-lines
2) Hand filleting line	11) Control weight
3) Filleting machine	12) Packing line
4) Filleting machine	13) Packing line
5) Skinning machine	14) Horizontal plate freezer
6) Skinning machine	15) Freezer tunnel
7) Skinning machine	16) Container for animal feed
8) Skinning machine	17) Container for animal feed
9) Drying transport	18) Washing machine for cut

19) Bone separator

A by-product is a product, produced by a manufacturing process, which adds a relatively small amount, if any, to the total market value of all product outputs.

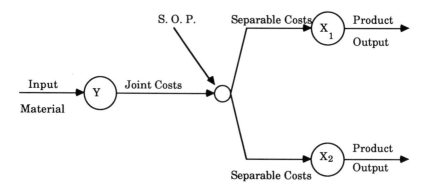

Fig.2 Manufacturing process.

The allocation of these joint costs to the finished products and the determination of which products, if any, are by-products has been discussed by various authors. (See especially Kaplan, 2.) The technique which will be used in this paper to allocate joint cost is mathematical programming (nonlinear programming). The primary questions which will be answered by the nonlinear model are: "How many tons/kilos of raw material should be bought and processed in a given case, and how to allocate the cost of raw materials and the overhead costs which occur before SOP?"

Attention should be given to one limitation when using Lagrange multipliers to allocate joint cost, that is the multiplier is only valid for very small changes from the optimizing values. The Lagrange multiplier (the joint cost allocation) will change continuously as the amount of any product produced is expanded [1,2].

Assumptions and Data

Following is an example which has certain assumptions. First of all this typical plant has unlimited raw material and processing capacity. For each unit of raw material input to the production process there can be several outputs or finished products (both regarding product types and number). As an example, consider a situation in which for 1 kg of raw fish (cod) as an input, 0.266 kg of fillets, 0.123 kg of block, 0.077 kg of minch, 0.525 kg of waste and 0.009 kg of overweight, are obtained as output.

Let's name the fillets = product-1 ("5 lbs")

- - - - block = product-2
- - - - minch = product-3
- - - - waste = product-4

Let: P_i = the price of product i, where i = 1...4 to indicate products 1, 2, 3 and 4.
  " : X_i = the volume of product i produced and sold.
The total yearly catch of cod in the North Atlantic is approximately 1,400,000,000 kg, and Iceland's part of this is around 300,000,000 kg or 21.4%. The plant in this example receives about 14,000,000 kg per year, or 4.7% of the total Icelandic cod catch. The reason for only looking at the cod catch in the North Atlantic is that this kind of fish creates its own special market. This is because of quality and processing methods. After considering these assumptions and what effect this particular plant could have on the market, the assumption was made that the following relations hold between the products volumes and prices (for 1 kg):

$P_1 = 220 - X_1 / 3,724,000$ in Ikr. (Icelandic krona)
$P_2 = 180 - X_2 / 1,722,000$ " "   (1)
$P_3 = 53 - X_3 / 1,078,000$ " "
$P_4 = 0$ " "

The 0.525 kg of waste does not have any price in this case, because it will be utilized for animal feed processing by a different company. So the company is at least free from the transport cost (price for such waste is usually very low, if any at all). Assume also that the raw material (raw cod) is bought at a price Ikr. 38./kg. The variable cost before the S.O.P. is Ikr. 30/kg of processed input material.

The Model

Let y be the number of kilos (kg) of the raw material to be processed. Then 0.266y kg of product-1, 0.123y kg of product- 2, 0.077y kg of product-3 and 0.534y kg of waste and overweight can be produced [5].
Then the model can be developed as:
Variables in the model: X_i = volume of product i produced.
                Y = volume of input material.
                P_i = price of product i.
The total revenue will be:
$$TR = (P_1)X_1 + (P_2)X_2 + (P_3)X_3 + (P_4)X_4 \qquad (2)$$
The results from solving this model (shown in the appendix) indicate that the company should buy 251,474 tons of raw material, and the optimal division of that volume between the finished products is:

Product-1 66,892 tons
  "   2 30,931 "
  "   3 19,364 "
  "   4 134,287 "

With regard to these volumes, the joint costs should be allocated in following manner:

to product-1 Ikr. 184.08 /kg-out
    "   2 "   144.08 " "
    "   3 "   17.08 " "
    "   4 "   0.00 " "

Furthermore, the model shows that when producing these volumes the price other than for completion of the product should be:

    product-1   202.04 Ikr./kg
    "   2   162.04 "   "
    "   3   35.04 "   "

All these values are only valid for small changes from the optimizing values. According to the recommended input-volume it would be theoretically the best choice to have only one big freezing plant in all Iceland. This depends of course on the overhead costs in this example, but is not currently practical [1,2]. However, this is the key usage of the method external to the actual firm.

DISCUSSION

Mathematical programming is useful in giving insight into the nature of the joint-product problems, but it also provides a basis for distinction between joint-products and by-products. If the value of Lagrange multiplier becomes $L_j < 0$, then the nonnegativity constraint has been violated and the solution is not correct. If this happens, then the company should produce less of the relevant product. A new solution has to be found where $L_j = 0$, and then this product must be a by-product and no joint-cost will be allocated to that product [1,2].

But in these solutions for the individuals firms in a national industry a set of capacity constraints would be used to limit the range of solution to the local (that is just one firm) optimum. By ignoring the capacity constraints, the non-linear model suggests that the best use of joint production facilities and raw material would lead a firm to increase drastically in size. When many such firms face similar raw material and pre-S.O.P. costs, a national policy can be deduced. Co-operative usage of common facilities or consolidation of buying would be moving the industry towards the optimum cost/price location. Where government intervention is desired in a national industry, it can be guided by the use of this technique.

Obviously, there are political and social questions beyond this proposed model. Any ability to add such concepts to the model would be a further advance for mixed economies especially in smaller countries where a national industry would have a large effect on the entire economy.

APPENDIX

Variables for the Model: X_i = volume of product i produced.
                                     Y   = volume of input material.
                                      P_i = price of product i.

The total revenue will be:

$$TR = (P_1)X_1 + (P_2)X_2 + (P_3)X_3 + (P_4)X_4 \tag{3}$$
$$TR = (220\text{-}X_1/3724000)X_1 + (180\text{-}X_2/1722000)X_2 + (53\text{-}X_3/1078000)X_3 + 0X_4 \tag{4}$$

Maximize the objective function:

$$Z = 220X_1 - X_1^2/3724000 + 180X_2 - X_2^2/1722000 + 53X_3 - X_3^2/1078000 + 0X_4 - 68Y \tag{5}$$

Subject to:	With Lagrange Multiplier:	
$X_1 \leq 0.266Y$	L_1	
$X_2 \leq 0.123Y$	L_2	(6)
$X_3 \leq 0.077Y$	L_3	
$X_4 \leq 0.534Y$	L_4	

or

Maximize:

$$Z = 220X_1 - X_1^2/3724000 + 180X_2 - X_2^2/1722000 + 53X_3 - X_3^2/1078000 + 0X_4 - 68Y +$$

$$L_1 (0.2766Y\text{-}X_1) + L_2 (0.123Y - X_2) + L_3 (0.077Y\text{-}X_3) + L_4 (0.534Y\text{-}X_4) \tag{7}$$

with all variables $\geq$ zero.

By taking the first order conditions:

$$\partial Z/\partial X_1 = 220 - 2X_1/3724000) - L_1 = 0 \qquad \partial Z/\partial X_2 = 180 - 2X_2/1722000) - L_2 = 0$$

$$\partial Z/\partial X_3 = 53\text{-} 2X_3/1078000) - L_3 = 0 \qquad \partial Z/\partial X_4 = \quad 0 - L_4 \quad = 0$$

$$\partial Z/\partial Y = -68 + 0.266L_1 + 0.123L_2 + 0.077L_3 + 0.534L_4 = 0 \tag{8}$$

$$\partial Z/\partial L_1 = 0.266Y - X_1 = 0 \qquad\qquad \partial Z/\partial L_2 = 0.123Y - X_2 = 0$$

$$\partial Z/\partial L_3 = 0.077Y - X_3 = 0 \qquad\qquad \partial Z/\partial L_4 = 0.534Y - X_4 = 0$$

Substituting the last four condistions into the first four:

$$L_1 = 220 - (2\bullet0.266Y/3724000) = 220 - (0.532Y/3724000)$$
$$L_2 = 180 - (2\bullet0.123Y/1722000) = 180 - (0.246Y/1722000)$$
$$L_3 = \quad 53 - (2\bullet0.077Y/1078000) = \quad 53 - (0.154Y/1078000) \tag{9}$$
$$L_4 = \quad 0$$

Substituting these into the $\partial Z/\partial Y$ equation:

$$\partial Z/\partial Y = -68 + 0.266(220 - (.532Y/3724000)) + 0.123(180\text{-}(.246Y/1722000))$$
$$+ 0.077(\ 53 - (.154Y/1078000)) + 0.534(0) = 0 \tag{10}$$

or   16.741 - 0.0000000665Y = 0

or   .0000000665Y = 16.741

and   Y = 251474250

Solving for the optimal output volumes:

X_1 = 0.266 • 251474250 = 66,892,151 kg or 66,892 tons (metric)

X_2 = 0.123 • 251474250 = 30,931,333 kg or 30,931 tons

X_3 = 0.077 • 251474250 = 19,363,517 kg or 19,364 tons

X_4 = 0.534 • 251474250 = 134,287,250 kg or 134,287 tons

Determining allocations of costs:

L_1 = 220 - (0.532 • 251474250/3724000) = 184.08 Ikr/kg output

L_2 = 180 - (0.246 • 251474250/1722000) = 144.08 Ikr/kg output

L_3 = 53 - (0.077 • 251474250/1078000) = 17.08 Ikr/kg output

L_4 =   0   =   0   Ikr/kg output

Allocation to product 1:   184.08 • 0.266   =   48.97 Ikr/unit

Allocation to product 2:   144.08 • 0.123   =   17.72 Ikr/unit

Allocation to product 3:   17.08 • 0.077   =   1.31 Ikr/unit

Allocation to product 4   0.00 • 0.534   =   0.00 Ikr/unit

              Total to be allocated   =   68.00 Ikr

Prices to the split-off point are:

Product 1: 220-(66892151/3724000) =   202.04 Ikr/kg output

Product 2: 180-(30931333/1722000) =   162.04 Ikr/kg output

Product 3:   53-(19363517/1078000) =   35.04 Ikr/kg output

Product 4:   0   =   0.00 Ikr/kg output

REFERENCES

1. F. Tayyari, Engineering Cost Analysis, Dept. of IE Bradley University, Peoria, Illinois, 1986.

2. R. S. Kaplan, Advanced Management Accounting, Prentice-Hall, Englewood Cliffs, NJ,1982.

3. D. E. Kroll & B. R. Weiss, Cost and Managerial Reporting Systems for Advanced Manufacturing Processes, presented to the "Tenth International Conference on Production Research", Nottingham, U.K., 1989.

4. Torbjorn Pedersen, Prosesser og Produkter i Norsk Fiskeindustri, Bind 2 -Del I, Universitetsforlaget, 1979.

5. Kristjan B. Gardarsson, Omleagning af Produktionen i en Filetfabrik, AUC,   Institut for Produktion, 1988.

Conceptual Model of an Activity-Based Cost Management System

Denise F. Jackson
University of Tennessee, Knoxville, TN 37996-1506

Thomas G. Greenwood
Carrier Corporation, Knoxville, TN

ABSTRACT

Technological advances in manufacturing industries have changed the processes involved in producing end-products, the elements within the processes, the end-products themselves, and, in turn, the management and control of these processes.

Cost and performance information which reflect the actual nature of operations in high-tech environments are vital for effective management. Such information would facilitate continuous process improvement, effective product costing, and proactive cost estimating.

This paper presents a model for a cost management system in such an environment. This system relates cost and performance information to the individual activity level and associates activities with the processes by which goods and services are designed, procured, produced, delivered and supported.

BACKGROUND

Much of what was performed by direct labor is now performed by automated machinery with the aid of computers. Thus, the contribution of direct labor is diminishing while that of indirect labor is steadily increasing. Also, nonvalue-added process activities are being identified and eliminated to streamline processes; products are being designed for manufacturability and maintainability; organizational lines are fading as more work is being performed by cross-functional teams; and management schemes are having to adapt to enable effective control in this new manufacturing environment.

Japanese organizations have already made great strides in adapting their management techniques to high-tech manufacturing environments. We have taken some of their experience into account in developing this model.

First, this model is not intended to be used to generate cost information for financial reporting or for the routine generation of production standard cost variances. This design decision reflects a conscious desire to follow the Japanese lead in severing the ties between process cost information and aggregate financial reporting data.

Secondly, this model has been developed to provide information detailed enough to support operational decisions regarding process improvement initiatives. This model was origionally designed to provide manufacturing cost data for cellular manufacturing environments with focused product offerings. In this aspect, our model differs from other current activity-based cost systems which are primarily oriented to the product cost application. It has been stated many times by Robin Cooper that the most successful implementations of activity-based cost systems to date have been in corporate environments where large product diversity created the opportunity for distortions in product cost assignments due to unit-based cost systems (Cooper, 1990). Our model provides an opportunity to extend activity-based costing techniques into industries with focused product offerings.

MODEL DESCRIPTION

This model consists of three distinct facets. The first facet develops the activity-based architecture of the existing organization. This architecture delineates organizational unit responsibilities, individual work activities, and the processes by which goods and services are provided. This phase clearly establishes the relationships among organizational units; the resources for which they are financially responsible; the activities which consume those resources; and the processes which sequence the activities to achieve a desired result, namely a product or a service.

The second facet entails the design of a cost assignment methodology to relate current expenses to the activities and processes that incur costs. Customer-related performance measures, such as lead time and quality information, are also recorded at the activity and process levels. This facet enables managers to conduct diagnostic assessments and benchmarking of current activity performance and to identify non-value added and high profile activities that will be targeted for improvement initiatives.

The third facet involves the development of a methodology for detailed process cost modeling and predictive analysis of alternative production scenarios. This methodology may be employed as a decision support tool to generate cost estimates for a wide array of operational decisions. Detailed descriptions of each phase follow.

Phase I: Activity-Based Architecture

In this first phase, the activity-based architecture is constructed around the existing organization and the process perspective is introduced to organizational managers. Processes are defined as the total network of activities required to produce a good or service or to accomplish a major business function. For the purpose of activity analysis, processes may be divided into smaller units which represent a specific component of the total activity network. These sub-processes may be further divided into components called process segments. Each process segment consists of a limited number of activities. In this manner, large systems may be reduced to the activity level for cost analysis. Figure 1 shows the hierarchy of work as defined within the process management orientation. For the purpose of analyzing cost and performance information, this model emphasizes analysis at the activity level and above in the hierarchy.

The significant result of this analysis is the creation of a multi-perspective view of the work activities within an organization. This view stresses the flow of products and/or information between activities without regard to organizational domains. Costs are analyzed from a holistic view which allows the cost optimization of the total process. This orientation suggests that cost improvements should be achieved through 1) the elimination of non-contributing activities and 2) the improvement of those activities that are truly critical to the overall performance of the total process.

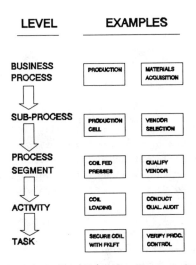

Figure 1. The Hierarchy of Work within the Process Orientation.

Phase II: Baseline Assessment

In order to establish a baseline assessment of current process level cost performance, operating costs will be allocated from the general ledger through organizational units that exercise budgetary control to individual activities as illustrated in Figure 2. Once cost information is available at the activity level, this information may be aggregated horizontally along process lines (where the process costs equal the sum of all the

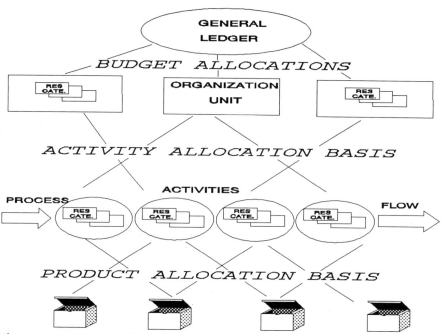

Figure 2.   Process-Oriented Cost Allocations.

activity costs) or further allocated vertically to the product level. The procedures for this facet of the model are outlined below:

1. From the general ledger accounts, the first step is to determine the ledger account codes that will be consolidated into each of the resource categories used to trace costs through organizational units to activities.   A resource category is defined as a collection of all the expenses associated with a generic resource type such as direct labor, utilities, equipment, tooling, etc.

2. Next, the general ledger expenses are allocated to organizational units as is traditionally done for budgetary purposes. Once a portion of the total resource category cost has been allocated to a given organizational unit, an index is affixed to the resource category code to associate that cost with the respective organizational unit that will control its disposition.

3. Costs are then allocated from the organizational unit level to individual activities, by resource category, using an activity allocation basis which is the contribution of each activity to the total consumption level of a given resource category. The resource consumption level is

reported in terms of the resource consumption basis, a unit measure for expressing the number of resource units that have been consumed.

4. Once the activity costs have been identified, activity costs may be rolled up to evaluate existing process cost performance. Also, costs may be aggregated to the process or sub-process level either by activity or by resource category.

5. The next step is to establish the performance measures, in addition to cost, that will be used to benchmark the process with respect to continuous improvement initiatives.

6. Next, two intermediate steps are performed. The first is to identify high profile activities with respect to excessive cost or poor performance. The second is to develop insights about how to improve the process functions that deal with the interactions between activities and to assess the value of individual activities in terms of their contribution to total process performance.

7. Current product costs may be determined at this point directly from the existing activity cost data by allocating costs using a product allocation basis as a third stage cost driver. (This step is necessary to provide an allocation from activities to products.) The product allocation basis is a percentage that apportions the cost of a specific activity to the products that create the demand for the activity. The product allocation basis is determined by a function or equation comprised of one or more cost drivers and/or product cost attributes.

Phase III: Process-Oriented Cost Modeler

The third facet includes a methodology for performing proactive cost modeling of alternative production scenarios. The specific steps are outlined below:

1. Explicitly define the objectives for developing the process cost model.

2. Define the model scope. Include the process segments and respective activities to be modeled, and specify the organizational units which provide resources in support of these activities. Also, identify specific product lines that place demands on the process segments.

3. Next, develop the cost and consumption profiles for each resource category. A profile is a graphical representation of the "short term" behavior of the cost or consumption of a given resource over an anticipated range of capacity levels. The consumption profile identifies the number of distinct resource units required to support various consumption scenarios. The cost profile identifies the cost of the resource for various capacity levels.

4. Prepare for the reconciliation process. The consumption requirements for a given resource category are the aggregate requirements from several different activities. These requirements are determined by

evaluating, by resource category, the required consumption level for each activity supported by the organizational unit, and then adding these consumption levels to get the aggregate requirements to be reconciled.

5. The cost driver relationships and cost/consumption profiles represent the cost generation logic for the simulation model. The next step is to validate the model against current baseline activity costs. Generate the estimated activity consumption levels based upon a specified product mix and known process parameter conditions. Then use the aggregate resource requirements from this step to enter the resource category cost and consumption profiles for each organizational unit.

Any discrepancies between the cost model output and the actual cost disposition should be resolved before continuing with the cost modeling process.

6. Once the initial cost model has been validated, evaluate the process costs of alternative production scenarios. First ascertain whether the new scenario will add, delete or change the nature of the activities that are inherent in the processes under evaluation; and adjust the activity-based system model as needed. Second, determine the product mix. Third, assign appropriate values to the product and process cost attributes used in the simulation model. Last, ascertain the impact of each alternative scenario on activity and process segment level performance measures. Then determine whether the result is closer or further away from the target performance goal.

7. Once the cost model variants have been constructed for each alternative production scenario, these models may be exercised to simulate different process costs by generating the resource consumption levels and reconciling the costs for each alternative.

The first action is to identify all the organizational units that support the process segments being modeled. Then, all the activities are identified that consume a given resource category for each organizational unit. Once the relevant activities are identified by resource category, the respective activity consumption levels are calculated as described in step four above and then aggregated as shown below to determine the total resource consumption level to be reconciled:

The next step is to determine, via the consumption profile, the number of whole resource units that will be required to meet demand requirements. The consumption profile consists of the resource consumption level, expressed in terms of the consumption basis, along the X-axis and discrete resource units along the Y-axis as shown in Figure 3. The procedure is to enter at the resource consumption level and move upward to intersect the profile line and across horizontally to the left to determine the number of whole resource units that are required. The example illustrated in Figure 3 reflects a step function consumption profile which is typical of situations in which resources are acquired in discrete units. Resource consumption

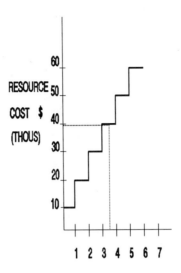

RESOURCE COST $ (THOUS)

60
50
40
30
20
10

1 2 3 4 5 6 7

RESOURCE CONSUMPTION LEVEL

(FRACTIONAL UNITS)

Figure 3. Resource Reconciliation: Cost Profile.

profiles may be fixed, linear, or step functions, depending on the nature of the resource category and the process situation.   The resource category cost for the reconciliation point is determined by entering the cost profile at the resource consumption level a shown in Figure 4. The procedure is the same as with the consumption profile; move upward to intersect the cost profile then horizontally to the left to determine the resource cost.

8.   During the resource reconciliation process the consumption and cost profiles are used to forecast the support organization's resource cost for each production scenario. Once this information is obtained, resource costs are allocated back from the resource reconciliation points to the individual activities for each resource category. The activity allocation basis is used to apportion these costs.

The calculation of the activity allocation basis may differ as the system scope changes. It may be determined by consensus estimates of demand requirements, or it might represent the actual proportion of the resource demand as calculated from the cost relationships specified in the

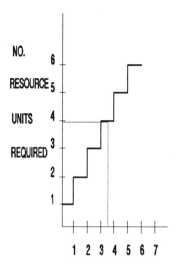

RESOURCE CONSUMPTION LEVEL

(FRACTIONAL UNITS)

Figure 4.Resource Reconciliation: Consumption Profile.

cost model.   With this methodology, the total cost of an activity may be determined by summing the cost of each resource category. Process costs are determined in a similar fashion by summing all the individual activity costs within the process either by aggregate totals or by individual resource categories.

9. Once the costs of activities and processes are established for each alternative scenario, this information may be used to evaluate operational decisions involving process cost alternatives, budget projections, and product cost estimating.

MODEL APPLICATION

The model should be applied in three phases. First, the activity-based architecture is developed for the entire organization in small companies or at the business unit level in large organizations. If necessary, the activity architecture may be developed incrementally by process.

This initial phase requires extensive interviews. The process-oriented perspectives that result will provide new visions and insights into how work is actually performed. This effort will also highlight the criticality of process level coordination to integrate individual activities for improved performance.

The activity-costing phase may also be implemented at the business unit level or at the first level at which a general expense ledger is maintained.

The third phase is to be conducted in an incremental fashion. Developing the prerequisite data structures to support cost modeling efforts is a complex task; therefore, the practitioner will probably begin by analyzing selected areas in support of specific decision requirements. As with other simulation techniques, once a model is developed for a given process segment, it can easily be modified and expanded to include new areas.

SUMMARY

In this paper, we have provided a description of a conceptual model for an activity-based cost management system. The foundation of this model is the hierarchical structuring of the work. This architecture connects resources consumed in activities to appropriate responsible organizations. It facilitates a downward dissemination of cost management decisions and an upward aggregation of cost and performance information.

Added to this foundation is a methodology for detailed process cost modeling and predictive analysis of alternative production scenarios. This feature goes a step beyond traditional cost management systems by not only meeting the process-control needs of lower and middle management, but also serving as a decision support tool to evaluate the impact of strategic decisions on the operations.

BIBLIOGRAPHY

Brimson, J. A. "Activity-based Investment Management" AMA Management Briefing, New York: AMA Membership Publications Division, 1989.

Conner, S. "Advanced Costing for Advanced Manufacturing." Proceedings of 1988 Conference: American Production and Inventory Control Society, 1988, pp. 567-570.

Cooper, R. "Cost Classification in Unit-Based and Activity-Based Manufacturing Cost System." Journal of Cost Management for the Manufacturing Industries, Fall 1990, pp. 4-14.

Cooper, R. and P. Turney. "Internally Focused Activity-Based Cost Systems." Proceedings of the Third Annual Management Accounting Symposium, San Diego, March 1990a, pp. 92-101.

Cooper, R. "The Rise of Activity-Based Costing--Part One: What Is an Activity-Based Cost System?" Cost Management Concepts and Principles, Summer 1988a, pp. 45-54.

Cooper, R. "The Rise of Activity-Based Costing--Part Two: When Do I Need an Activity-Based Cost System?" Cost Management Concepts and Principles, Fall 1988b, pp. 41-48.

Cooper, R. "The Rise of Activity-Based Costing--Part Three: How Many Cost Drivers Do You Need, and How Do You Select Them?" Cost Management Concepts and Principles, Winter 1989a, pp. 34-46.

Cooper, R. "The Rise of Activity-Based Costing--Part Four: What Do Activity-Based Cost Systems Look Like?" Cost Management Concepts and Principles, Spring 1989b, pp. 38-49.

Dugdale, D. and S. Shrimpton. "Product Costing in a JIT Environment." Management Accounting: Journal of the Institute of Cost and Works Accountants, March 1990, pp. 40-42.

Dunn, A. "How to Determine the Real Cost of your Products." Proceedings of 1988 Conference: American Production and Inventory Control Society, 1988, pp. 536-541.

Greenwood, T. An Activity-Based Conceptual Model for Evaluating Process Cost Information, Doctoral Dissertation, University of Tennessee, May 1991.

Greenwood, T. "Budget Planning Analysis." Unpublished research findings, University of Tennessee Research Contract R01136063, June 1989.

Huthwaite, B. "The Link Between Design and Activity-Based Accounting." Manufacturing Systems, October 1989, pp. 43-47.

Janson, P. and M. E. Bovarnick "How to Conduct a Diagnostic Activity Analysis: Five Steps to a More Effective Organization." National Productivity Review, Spring 1988, pp. 152-160.

Jeans, M. "The Practicalities of Using Activity-Based Costing." Management Accounting: Journal of the Institute of Cost and Works Accountants, November 1989, pp. 42-44.

Johnson, G. and M. Stevens "Cost Accounting in the Factory of the Future." Proceedings of 1988 Conference: American Production and Inventory Control Society, 1988, pp. 571-573.

Johnson, H. et al. "Activity Management and Performance Measurement in a Service Organization." Proceedings of 1989 Symposium: Management Accounting, 1989, pp. 63-73.

Kaplan, R. "Management Accounting for Advanced Technological Environments." *Science*, Vol. 245, August 1989a, pp. 819-823.

McNair, C. J. "Interdependence and Control: Traditional vs. Activity-Based Responsibility Accounting." *Journal of Cost Management for the Manufacturing Industry*, Summer 1990, pp. 15-25.

Monden, Y. and M. Sakurai. *Japanese Management Accounting: A World Class Approach to Profit Management*, Cambridge: Productivity Press, 1990.

Patell, J. "Adapting a Cost Accounting System to Just-in-Time Manufacturing: The Hewlett-Packard Personal Office Computer Division." *Accounting and Management*, Fall 1989, pp. 229-267.

Prather, K. "Cost Accounting For CIM/JIT/MRP." *American Production and Inventory Control Society 1988 Conference Proceedings*, 1988, pp. 533-535.

Rotch, W. "Activity-Based Costing in Service Industries." *Journal of Cost Management for the Manufacturing Industry*, Summer 1990, pp. 4-14.

Turney, P. "Using Activity-Based Costing to Achieve Manufacturing Excellence." *Cost Management*, Summer 1989.

Waigh, M. and R. Hunt "Cost Accounting in a JIT Environment." *Proceedings of 1988 Conference: American Production and Inventory Control Society*, 1988, pp. 548-549.

Yoshikawa, T. et al. "Japanese Management Accounting: A Comparative Survey." *Management Accounting: Journal of the Institute of Costs and Works Accounts*, November 1989, pp. 20-23.

VI. Cost Sensitivity Analysis

Sensitivity Analysis Using Discrete Simulation Models

*Alan S. Goldfarb, Arlene R. Wusterbarth, Patricia A. Massimini,
and Douglas M. Medville*
The MITRE Corporation, 7525 Colshire Drive, McLean, VA 22102

Introduction

Simulation models are commonly constructed to permit the study of the behavior of a system when it may not be feasible or practical to study the real system, e.g. in the early stages of design and development of a complex system. It may, for example, be desirable to evaluate the effect on system performance of variations in subsystem performance parameters, operating conditions, or design configurations. It is frequently desirable to determine the sensitivity of total system performance to the performance of its individual subsystems so that subsequent testing and design improvement efforts can concentrate on those subsystems that have the greatest effect on system performance. Thus, the cost of extensive testing of those subsystems that have a lesser impact on total system performance can be avoided.

The traditional approach to sensitivity analysis usually involves a factorial experiment. In this type of experiment, a set of values is established for each parameter (factor) of the subsystems that make up the system under study. Examples of parameters include the means of exponential distributions that characterize subsystem failure or repair times. One or more simulations are carried out with each of the possible combinations of the values of the factors. The output of each simulation run is an estimate of the total system performance when the factors are characterized by the particular set of values used in the run. A systematic procedure is used to analyze and interpret the results of the experiment (for example, see Davies 1960). Unfortunately, when the number of factors to be investigated becomes large (e.g. greater than 9), the number of simulation model runs required for analysis becomes prohibitively large. For example, a full factorial experiment for 22 factors, each with two values, would require 4,194,304 simulation runs.

An alternate procedure, described by Iman, Heltor and Campbell (July 1981; October 1981), allows the analyst to capture the influence of many factors on the performance of the modeled system without requiring an excessive number of computer model runs. The purpose of this paper is to illustrate the use of this procedure with data generated from a series of discrete simulation model runs and to demonstrate its usefulness in highlighting a small set of important parameters.

Description of Modeled System

A complex operating system is usually made up of numerous subsystems connected in series and/or in parallel and operated in a stochastic manner. An example of such a system is depicted in Figure 1. This system might, for example, represent a plant for the assembly of some type of commercial item. The illustrated system consists of 7 subsystems, several of which are duplicated (e.g., Subsystem 1A is identical to Subsystem 1B). Subsystems 1A and 2A operate in parallel with Subsystems 1B and 2B. Subsystem 3 must be operating for Subsystem 2 to operate (e.g., Subsystem 3 may be a hydraulic system that provides power to Subsystem 2). Subsystem 5 is fed by Subsystems 2A and 2B alternately when both are operating, otherwise, Subsystem 5 is fed only by the operating subsystem. In either case, the feed rate to Subsystem 5 is the same as long as either Subsystem 2A or 2B is operating. Subsystem 5 produces the desired product and a byproduct which goes into Buffer 1 (a storage area). Subsystem 2 produces two outputs: one goes to Subsystem 5, as previously discussed, and the other goes into Buffer 2. This buffer accumulates items from both Subsystems 2A and 2B. The items in Buffer 2 are processed by Subsystem 4 and the product, which is similar to the byproduct from Subsystem 5, goes into Buffer 1. Buffer 1 feeds Subsystem 6, which in turn feeds Subsystem 7.

The use of buffers between subsystems allows continued operation of one of the subsystems when the other is down. For example, if Buffer 2 contains some items from Subsystem 2, Subsystem 4 can continue to operate when Subsystems 2A and 2B are shut down for repair. Similarly, if Buffer 2 is not full, and Subsystem 4 is down for repair, Subsystems 2A and 2B can continue to operate and feed Buffer 2.

For this system, the performance characteristic of interest is the rate (R) at which the product will be produced. This rate is dependent on the rate at which each subsystem operates and/or the amount of time that each subsystem is down for repairs. The latter is a function of the frequency of breakdowns and the average repair time for the subsystem. The objective of the sensitivity analysis is, therefore, to determine which variables (operating rate, frequency of failure, or repair time) associated with which subsystems are critical to attaining the desired system performance.

Simulation Methodology

The computer simulation model of the system depicted in Figure 1 was written in a personal computer (PC) version of General Purpose Simulation System (GPSS) language developed by MINUTEMAN SOFTWARE. GPSS/PC™ is particularly well suited for modeling this type of a system because it is designed for "reproducing the dynamic behavior of systems which operate in time, and in which changes of state occur at discrete points in time" (Schriber 1974). In this system, individual items spend a discrete amount of time in each subsystem before moving to the next subsystem for subsequent processing.

In order to model the variation expected during normal operation, the model is constructed as a stochastic process with respect to subsystem failure and repair time. However, it is deterministic with respect to the processing time in each subsystem. Subsystem failures may be cycle dependent, time dependent, or both. Cycle dependent failures occur after some number of items have been processed. Time dependent failures occur after some period of time has elapsed following the start of operation.

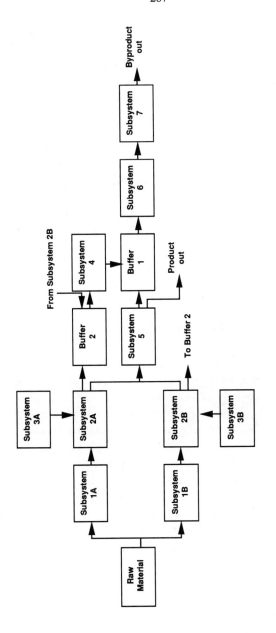

Figure 1. Block Diagram of a Complex System

During a simulation, the model selects values of the times or cycles to failure and times to repair randomly from an assumed probability distribution function. For the modeled system, exponential probability distributions were assumed for all times to fail, cycles to fail, and times to repair. Exceptions to this were for Subsystems 4 and 5 where the time to repair was assumed to be lognormally distributed. The exponential distribution requires only a single parameter (the mean) to be fully specified. The lognormal distribution requires two parameters, the median and standard deviation, to be fully specified.

Before using the computer simulation model for examining the behavior of the system, it is necessary to first establish confidence that the model adequately represents the system. Confidence in the model was established by confirming that the internal logic functions the way it was intended. This was accomplished by using the various features of GPSS that allow the behavior of the simulation to be visually observed on active graphics screens. GPSS/PC^{TM} also generates a standard report that provides an additional means of verifying that the model is behaving as intended.

The output of the simulation model is the total quantity of product produced during the simulated production period. It is important to select a simulated time so that two competing goals can be satisfied: (1) the time should be long enough to provide statistically significant information about the production rate; and (2) the time should be as short as possible to minimize computer time so that a large number of repetitions for different parameter values could be performed.

In order to develop a quantitative basis for selecting the total simulation time, daily production quantities were generated by the simulation model for various lengths of operating time. A statistical analysis of the data indicated that the mean value of the daily production did not change significantly as the operating time increased from a period of two to ten 120-hour weeks. In fact, the largest and smallest values over this range of time periods differed by only 2 percent, implying that a short simulated operating time period could be justified. Moreover, for a simulated operating time of three weeks, the endpoints for the 90 percent confidence interval for the values of daily production were within +/- 5 percent of the mean value for daily production. It was judged that the slight increase in precision attained by increasing the simulated operating time was not warranted by the additional running time (about 15 minutes for each simulated week of system operation using an IBM PC/AT^{TM}) that would be required for the simulation, and 3 weeks (360 hours) were selected as the simulated time period.

Analytical Approach

The goal of the sensitivity analysis is to characterize the response of system performance, in terms of production rate (R), to changes in the value of the system parameters (V_i). In this case, the system parameters are the processing rates (Rate) and the parameters of the probability distribution functions for the times or cycles between failure (MTBF or MCBF) and the times to repair (MTTR) for each subsystem. The sensitivity analysis was performed as follows:

(1)   The value of each of the 22 system variables (i = 1,2,...,22) was systematically changed, and the simulation model was run with the perturbed values to generate a value for the system

average production rate over the simulated time period. This procedure was repeated 49 times. The variation in the average system throughput between simulation runs is primarily due to the different values of the parameters used in each run.

(2)   The production rate was then regressed against the system parameters in a stepwise manner to identify the parameters which had the most influence on the value of the production rate.

The values of the 22 parameters for each of the 50 simulation runs were generated in two stages:

(1)   Probability distributions were developed for each of the 22 parameters. These can be based on prior knowledge for similar systems or equipment, preliminary tests, or engineering judgment. In this case, a triangular distribution was assumed for the distribution of possible values for the variables (see Figure 2). Analyses of similar systems indicate that the results are relatively insensitive to the shape of the assumed distribution, but that the range of values is important (Alpert et al. 1985).

(2)   A standard Monte Carlo procedure was used to randomly select specific values for the parameters from each of the 22 distributions. Fifty sets of 22 parameters, one for each simulation run, were selected by this procedure. It was assumed that there were no correlations among any of the parameters.

The decision to use 50 simulation runs for the sensitivity analysis was based on a survey of other sensitivity analyses with large computer models. The goal in these types of studies is to generate a distribution of model outputs so that they can be regressed against the values of the input variables. These other sensitivity analyses concluded that approximately 50 runs are adequate to characterize the output distribution. One study, which involved a comparable number of parameters, generated two sets of results: one based on 50 computer runs and one based on 100 runs. It was found that the results of the two experiments were very similar, and that, for the purpose of the sensitivity analysis, no more than 50 runs were required (Alpert et al. 1985).

The stepwise regression procedure involves constructing a sequence of regression models by adding one parameter at a time to the previous regression model. The regression equation used to fit the data was of the following form:

$$R = A_0 + \sum_{i=1}^{n} A_i V_i$$

where n is the number of parameters used in the regression model. The maximum value of n is 22. The parameter which gives the highest simple correlation with the production rate is selected as the first parameter in the regression model. Next, the parameter with the highest partial correlation is added to the regression equation. The process continues until all statistically significant parameters have been added.

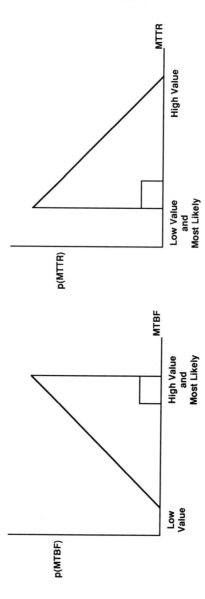

Figure 2. Triangular Distributions for Subsystem's Variables

Each time a parameter is added, a check is made to see if the significance of parameters already in the model has changed. If a parameter becomes insignificant, it is dropped from the regression equation (although it may reappear in a later step as other parameters are added and deleted). Ultimately, the technique produces the subset of parameters from among the available independent parameters that best describes the observed variation in the production rate (Draper and Smith 1966; Iman and Conover 1982). In this analysis the relative importance of the component parameters is determined by their contribution to the coefficient of determination (R^2) of the regression.

Example Sensitivity Analysis

A sensitivity analysis was performed for the system shown in Figure 1. The parameters of the triangular probability distribution functions used for the 22 variables are listed in Table 1.

A standard Monte Carlo technique was used to generate 50 random values from the probability distribution function of each of the parameters. Sample histograms of the values selected for three of the variables are illustrated in Figure 3.

A histogram of the production rates obtained from the 50 simulation model runs with the sets of randomly selected values is shown in Figure 4. The observed production rate ranges between 24.7 and 38.3 items per hour. The mean is 33.3 items per hour, and the standard deviation is 2.68 items per hour. This suggests that the simulation results are not widely dispersed.

Table 1
Parameters of the Probability Distribution Functions
for the System Variables

Parameter Name	Low Value	Most Likely Value	High Value
$MCBF_1$	1,128.	12,248.	12,248.
$MTTR_{1C}$	1.86	1.86	2.51
$MTBF_1$	27.62	299.13	299.13
$MTTR_{1T}$	1.86	1.86	2.10
$MCBF_2$	117.	887.	887.
$MTTR_{2C}$	2.80	2.80	3.34
$MTBF_2$	16.39	122.91	122.91
$MTTR_{2T}$	2.11	2.11	2.80
$RATE_2$	50.	60.	60.
$MTBF_3$	11.85	499.09	499.09
$MTTR_3$	2.20	2.20	2.75
$MTBF_4$	6.10	25.67	25.67
$MTTR_4$	1.29*	1.29*	4.24*
$RATE_4$	62.7	70.00	98.00
$MTBF_5$	6.10	25.67	25.67
$MTTR_5$	1.29*	1.29*	4.24*
$RATE_5$	39.6	47.00	50.00
$MTBF_6$	2,567.4	6,037.20	6,037.20
$MTTR_6$	2.57	2.57	6.04
$MTBF_7$	867.64	1,464.70	1,464.70
$MTTR_7$	4.36	4.36	7.36
$RATE_7$	42.00	70.00	70.00

----

*Medians of a lognormal distribution; it was assumed that the standard deviation was not a variable.

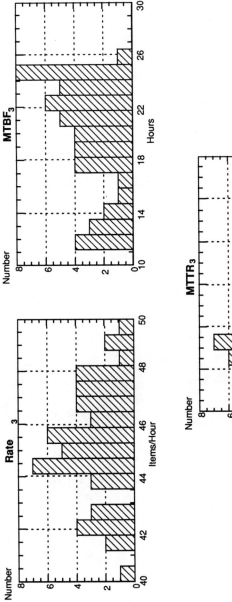

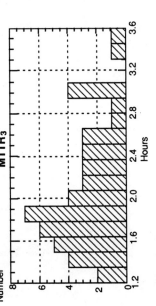

Figure 3. Random Samples for Selected Variables of the Example System

Number of Occurrences

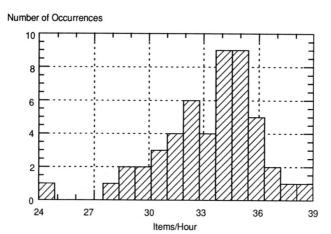

Figure 4. Frequency Distribution of System's Production Rate from Simulation Model

The system production rate was regressed against the system variables. The results of the stepwise regression analysis are presented in Table 2. Of the 22 system variables investigated, 7 were found to be statistically significant at the 5 percent level. The coefficient of determination shows that the regression equation explains 85 percent of the variation in the production rate. The system performance is seen to be most sensitive to the performance of Subsystem 5 in which the parameters for the distributions of processing rate, time between failures, and time to repair account for 91.7 percent of the total coefficient of determination of the regression equation.

Table 2
Importance Ranking of System Parameters

Parameter	Multiple Regression Coefficient	t-Statistic	Contribution to R^2 (Percent)
$RATE_5$	0.7357	10.7225	38.9
$MTTR_5$	-3.0882	-10.4539	32.8
$MTBF_5$	0.2100	5.4981	20.0
$MCBF_2$	0.0034	3.4120	4.6
$MTBF_1$	0.0049	2.2414	1.9
$MTBF_2$	0.0258	4.1500	1.7
$MTBF_3$	0.0062	4.3587	0.03

Note: A parameter is considered statistically significant if the absolute value of its t-statistic is greater than 2.02 (for a level of significance of 0.05 and 42 degrees of freedom).

The scatter diagram in Figure 5 illustrates the prediction capability of the regression equation. The points on the diagram represent the value of the production rate obtained from the simulation model (y-axis) and the value of the production rate predicted by the regression equation (x-axis) for the same set of parameter values. If the values are the same, the point falls on the straight line. As the figure shows, the regression equation replicates the 50 values generated from the simulation model quite well. Consequently, if an analyst wanted to know the effects of particular changes in any of the seven significant parameters of the system, the analyst could use the multiple regression equation as a convenient alternative to the simulation model. However, any changes made to the parameters should remain within the bounds of the values of the parameters used to develop the regression equation.

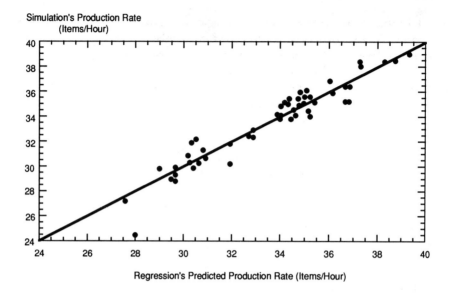

Figure 5. Prediction Capability of Multiple Regression Equation

References

Alpert, D.J., R.L. Iman, J.D. Helton, and J.D. Johnson, 1985. "A Demonstration Uncertainty/Sensitivity Analysis Using the Health and Economic Consequence Model CRAC2." NUREG/CR-4199. U.S. Nuclear Regulatory Commission, Washington, D.C.

Davies, O.L. 1960. The Design and Analysis of Industrial Experiments. Hafner Publishing Company, New York, N.Y.

Draper, N.R. and H. Smith, 1966. Applied Regression Analysis. John Wiley and Sons, New York, N.Y.

Goldfarb, A.S., A.R. Wusterbarth, P.S. Abel, and D.M. Medville, 1988. "Sensitivity Analysis Using Discrete Simulation Models." Paper presented at the 1988 Summer Computer Simulation Conference, Seattle, Washington.

Iman, R.L., J.C. Helton, and J.E. Campbell, 1981. "An Approach to Sensitivity Analysis of Computer Models: Part I - Introduction, Input Variable Selection and Preliminary Variable Assessment." Journal of Quality Technology, XIII-3 (July): 174-183.

Iman, R.L., J.C. Helton, and J.E. Campbell, 1981. "An Approach to Sensitivity Analysis of Computer Models: Part II - Ranking of Input Variables, Response Surface Validation Distribution Effect and Technique Variables Synopsis." Journal of Quality Technology, XIII - 4 (October): 232-240.

Iman, R.L. and W.J. Conover, 1982. "Sensitivity Analysis Techniques: Self-Teaching Curriculum," NUREG/CR-2350. U.S. Nuclear Regulatory Commission, Washington, D.C.

Schriver, T.J., 1974. Simulation Using GPSS. John Wiley and Sons, New York, N.Y.